Carbon Nanostructures

Carbon is intimately connected to almost everything we deal with in a daily basis. Due to its outstanding properties, such as high stability at environmental conditions, different hybridizations, strong covalent bond formation and easy of compounds formation, carbon has been a topic of scientific interest in several areas. Indeed, starting in the 19th century, chemists have devoted a whole field to study carbon-based compounds, which is, nowadays, known as Organic Chemistry. Remarkably, the last 30 years have been witnessing an exponential advance in the science involving carbon and carbon structures. Since the discovery of Fullerenes in 1985, which was awarded the Nobel Prize in Chemistry in 1996, carbon nanostructures have been attracting a great deal of attention from the research community. This public interest dramatically increased after the publications by the Iijima and Bethune groups on single-wall carbon nanotubes in 1993 and found a "new research era" with the isolation of a monolayer of carbon atoms, also called graphene, which conducted groundbreaking experiments demonstrating outstanding phenomena such as the Klein-Tunneling and the fractional quantum hall effect. No wonder, graphene was the object of the 2010 Nobel Prize in Physics.

The "Carbon Nanostructures" book series covers the state-of-art in the research of nanocarbons and their applications. Topics related to carbon allotropes such as diamond, graphite, graphene, fullerenes, metallofullerenes, solid C60, bucky onions, foams, nanotubes and nanocones, including history, theory, synthesis, chemistry & physics, Biophysics & engineering, characterization methods, properties and applications are welcome. Within the "Carbon Nanostructures" book series, the reader will find valuable, up-to-date account of both the newer and traditional forms of carbon. This book series supports the rapid expansion of this field and is a valuable resource for students and professionals in several different areas.

Springer welcomes new book ideas from potential authors. If you are interested in publishing your book in this series, please contact us via mayra.castro@springer.com

**** Indexed by Scopus (2018) ****

More information about this series at http://www.springer.com/series/8633

Santosh K. Tiwari · Sumanta Sahoo · Nannan Wang
Editors

Electrospinning of Graphene

Springer

Editors
Santosh K. Tiwari
School of Resources, Environment and Materials
Guangxi University
Nanning, China

Sumanta Sahoo
Department of Chemistry
Madanapalle Institute of Technology and Science
Madanapalle, Andhra Pradesh, India

Nannan Wang
School of Resources, Environment and Materials
Guangxi University
Nanning, China

ISSN 2191-3005 ISSN 2191-3013 (electronic)
Carbon Nanostructures
ISBN 978-3-030-75455-6 ISBN 978-3-030-75456-3 (eBook)
https://doi.org/10.1007/978-3-030-75456-3

This Springer imprint is published by the registered company Springer Nature Switzerland AG
The registered company address is: Gewerbestrasse 11, 6330 Cham, Switzerland

Preface

Organic materials are the center of our present civilizations, and several modern industries are using carbon materials in different forms (like food items, polymers, drugs, oils, etc.) to make our lives full of convenient. At present, nanotechnology is one of the most growing research fields and many exceptional innovations have been created using nanomaterials for the betterment of our lives. Carbon nanomaterial exhibition is enthralling nanoscale architectures, like hierarchical association, regular pattern, and several unique morphologies, which allows them to achieve high porosity and ultra-high surface area. At present, nanofibers with enormously porous mesh and their huge surface-to-volume ratio advance performance for the plentiful applications have shaped enormous attention of the scientific communities. Therefore, several new procedures and machineries were introduced time to time for the fabrication of nanofibers with the desired properties. The electrospinning technique has tremendous competence to produce nanofibers of diverse materials in countless fibrous assemblies. The scalable production of nanofibers and simplicity of the experimental setup make electrospinning very attractive for both academia and industries. A variety of nanofibers and analogs has been made for biomedical engineering, energy storage, environmental engineering, defense, and security applications. The auxiliary benefit of this technique is that it could produce desired types of nanofibers using variety of materials with required functional groups unto the surface of fiber by tuning the parameters of electrospinning device. This technique offers an adorable route to incorporate 2D nanomaterials, especially graphene and graphene oxide into the polymeric fibers. Thus, easy processing, high aspect ratio, processability with wide range of polymers, and straightforwardness make this practice very attractive for making unique architectures for miscellaneous applications. On the other hand, since the last two decades, graphene is dominating almost all research fields. In this book, we have presented an overview of electrospinning of graphene. The first chapter deals with the recent progress in this electrospinning technologies for fabrication of graphene-based materials. The second chapter describes the fabrication and application of graphene-infused polymeric flexible, stretchable and transparent devices. The next chapter summarizes the utilization of graphene nanofibers synthesized through electrospinning technique as fillers for dimension-controlled polymer matrices. Fourth chapter gives a brief overview on the supercapacitor application of graphene-based nanocomposites. Further, fifth chapter illustrates the battery application of graphene nanofiber and polymer composites. Lastly, the sixth chapter deals with the fuel cell application of graphene nanofiber-based composites. Overall, the book deals with the different aspects of electrospinning technology, specially focused on the fabrication of porous graphene materials. The fundamentals of electrospinning technology of graphene and their applications in different research fields have been summarized in this book. We hope this book will be highly beneficial for the researchers working in this field. Moreover, each chapter of this book has plentiful references associated with the electrospinning of graphene, which enable further interpretation and exploration.

Nanning, China — Santosh K. Tiwari
Madanapalle, India — Sumanta Sahoo
Nanning, China — Nannan Wang

Contents

Recent Progress in Electrospinning Technologies for Graphene-Based Materials

Shrabani De, Sumanta Sahoo, Ashok Kumar Das, and Ganesh Chandra Nayak

Abstract Electrospinning has emerged as a versatile and promising technique to synthesize nanofibres. With increasing demand of nanotechnology, electrospinning has gained more attention due to its versatile application in various fields. Scientists have incorporated various nanomaterials as nanofillers in the polymeric matrix to enhance the properties of nanofibres according to their specific applications. Among these nanofillers, graphene has gained extensive interest for researchers, as a multi-functional molecule associating different unique properties like high mechanical strength, electrical conductivity, flexibility, conductivity and optical transparency. These desirable properties make graphene a superior material than CNTs and other conducting nanoparticles. The graphene-based polymeric nanofibres have opened new opportunities for diverse applications of nanofibres in different walks of life. This chapter aims to describe an overview of progress of graphene-based electrospun nanofibres and their applications in various fields including biomedical, chemical, defence and environmental applications. The historical overview and fundamentals of electrospinning, graphene and its properties as nanofiller as well as the applications of graphene-based electrospun nanofibres in different fields are discussed. The limitations and future developments of electrospinning and graphene-based electrospun nanofibres that can be made are also presented.

Keywords Electrospinning · Polymer matrix · Graphene · Nanofillers · Electrospun nanofibres

S. De · G. C. Nayak (✉)
Department of Chemistry, Madanapalle Institute of Technology and Science, Madanapalle 517325, Andhra Pradesh, India
e-mail: gcnayak@iitism.ac.in

S. Sahoo (✉) · A. K. Das
Department of Chemistry, IIT(ISM) Dhanbad, Dhanbad 826004, Jharkhand, India

S. K. Tiwari et al. (eds.), *Electrospinning of Graphene*, Carbon Nanostructures,
https://doi.org/10.1007/978-3-030-75456-3_1

1 Introduction

Technological progress over the previous decades has been focused on different competing methods to synthesize nano-sized materials. The development of nanotechnology shows its excellence in selected fields, while electrospinning (E-spin) has been come out as a popular nanotechnology for the easy fabrication of nanofibres from wide range of materials. E-spin technique that proved to be an efficient method to produce electrostatic fibre with diameter around 2 nm to several micrometres using electrical force from the polymer solution of synthetic as well as natural polymers, has been attracted tremendous attention in research and commercial field over the past few decades due to the low cost production, chemical versatility, tunable physical properties like fibre length, porosity as well as electronic, optical, biomedical and mechanical property [1–3]. This process is widely used because of its unique capability and versatility of producing fibre and fabrics within nanorange consistently with controlled pore size which is difficult to obtain in standard mechanical fibre-spinning techniques [4–6]. From the starting of this century, researchers have been paying much attention on re-examination of E-spin technique [7]. Due to having high surface area than regular fibres, the fibres generated from E-spin have been successfully applied in wide range of fields including biomedical, tissue engineering, pharmaceutical, optical, defence, protective clothing, environmental engineering, electronics, healthcare and biotechnology [8–11]. Overall, E-spin is a simple and robust technique to produce fibre from wide variety of polymers, which offers a lot of advantages like tuneable pore size, high surface-to-volume ratio, malleability to generate a variety of shapes and sizes and the capability to generate controlled composition, property and functionality of nanofibres. Because of these properties, electrospun fibres have drawn major attention in the field of applied physics, chemistry, regenerative medicine, sensing, biological scaffolds, photonics, flexible electronics and electrode materials in energy storage application [12]. Over 60 years, E-spin technique has been applied in textile industry for the production of nonwoven fabrics. From the past few years, there has been increasing attention in the application of this technique for the manufacturing of nanolevel fibres from a range of natural and synthetic polymers used in tissue engineering [13] including cellulose [14], collagen [15], polyurethanes [16], cellulose/collagen [17], silk fibroin [18], and polylactic acid [19]. Although having various superiorities, the scale up of the output of the nanofibres suffers some serious problems which limit its bulk application. Several researches have been carried out on amplifying the production rate of electrospun fibres where a two-layer spinning system is used in which the lower layer with ferromagnetic suspension and the upper layer with polymer solution, and multiple nozzle or spinnerets systems are arranged in a circle/line/matrix and a bottom up gas jet spinning or bubble E-spin [6, 20, 21]. To increase the production of electrospun fibres using single jet is not very viable for large scale applications. Although porous hollow tube is used in various cases to obtain multiple jets and the production rates can be amplified by increasing the number of holes and tube length [22, 23]. Another major problem in tissue engineering with electrospun scaffolds and mats is the insufficient cellular migration and non-uniform cellular distribution in the scaffold under passive seeding conditions with enhancing depth.

However, using conventional E-spin technique, nanofibres can be synthesized in a simple and inexpensive way, but over the time, there are build-ups of meshes with very high fibre density. It is also reported that, when the diameter of fibres is smaller, the chance of fibre to fibre contact per unit length of fibre increases and the average pore radius in the mesh decreases [24]. Due to these circumstances, there occurs a huge size mismatch between the smaller pores in the fabrics and the larger size of the cells. This limits the ability of the cells to populate and emigrate inside the scaffolds. This problem limits the application of electrospun fibres, especially in the field of 3D organs or tissues. Among several attempts which addresses this limitation, Ekaputra et al. suggested three methods for the advancement of cellular infiltration and their viability [25]. Among the three methods, the first one was co-E-spin, in which a medical grade poly (Ɛ-caprolactone)/collagen (mPCL/col) was used as a main fibre combined with water soluble polymer gelatine and polyethylene oxide to enhance the pore volume by selective elimination of solid materials from the mesh. The approach of blending water soluble polymer didn't show enough improvement compared with conventional electrospun fibres used in cell infiltration. In the second approach, micron-sized mPCL/col was used to improve the fibre–fibre distance as well as pore volume. In the third approach, mPCL/col was co-deposited with heprasil (glycosaminoglycan hydrogel) which create enzymatically degradable pockets within the dense fibres through which cell can migrate. Mechanical booster of fibres through the introduction of nanofillers which enhances the property of fibres in particular field of applications. Incorporation of the nanomaterials into the matrix enhances to optical, mechanical, electrical and thermal properties of electrospun fibres. For the production of conducting nanofibre, the nanofillers can be classified into zero dimensional (0D), one dimensional (1D), two-dimensional (2D) and three dimensional (3D), which were introduced in this decade. Among various nanofillers, 2D layered graphene and graphene oxide has been attracted huge attention as they enhance the fibre properties due to their mechanical strength, thermal and electrical conductivities [26].

Around this decade, the discovery of graphene drastically changed the research direction of material science by its unique and splendid characteristics. All over the world, researchers are competing to discover different synthesis pathways and explore the applications of graphene-based materials in various fields. An intermediate material called graphene oxide (GO) associated with better hydrophilicity and chemical versatility with easier modification due to the presence of oxygen functionalities compare than that of graphene [27]. Over the years, a broad range of applications of graphene has been published such as tissue engineering, electronics, optics, energy storage and sensing [27, 28]. However, graphene has an unavoidable tendency to agglomerate due to the Van der waals force and $\pi - \pi$ stacking which results the loss of high surface area and excellent electric property. To prevent the agglomeration of graphene sheets, development of preferable scheme to synthesize fully dispersed graphene-based nanocomposites remains a significant challenge

[29]. GO-based materials have been introduced in nanofibre composites using E-spin technique. Incorporation of graphene in nanofibres enhances thermal and electrical conductivity, catalytic activity, mechanical strength and optical property.

In this chapter, we focus on comprehensive overview on recent advancement of E-spin designs and process and graphene-based electrospun nanofibres. It includes the enhancement of properties due to the incorporation of graphene into the fibre matrix and their various applications in selected fields. In this aspect, we summarize the basic principle and instrumentation of E-spin technique as well as application of graphene-based nanofibre composites in several fields including biomedical, chemical, defence and environment. The limitations and the future aspect of graphene-based nanocomposites fabricated by E-spin are also included.

2 Historical Overview of Electrospinning

E-spin was first distinguished by Rayleigh in 1897 [30]. The detailed study on electrospraying was reported by Zeleny in 1914 by establishing a mathematical model to explore the effect of electrical forces in liquids by analysing the behaviour of liquid droplets at the end point of iron capillaries [31]. In 1934, Formhals fruitfully fabricated electrospun fibres with an improved E-spin device and patented his work [32]. The ground work of E-spin was set by Taylor in 1969 through his work on electrically controlled jets [33]. The phrase "E-spin", acquired from electrostatic spinning was used in around 1994, which is comparatively recent, but its beginning was uncovered back to above 60 years ago. A series of patents was issued by Formhals, explaining the experimental setup for the synthesis of polymer filaments applying an electrostatic force [34]. The first patent (US Patent Number: 2116942) on E-spin technique was published for the synthesis of textile yarns using a voltage of 57 kV for the E-spin of cellulose acetate and acetone and monomethyl ether of ethylene glycol was used as solvents. The patent holder of this process was Antonin Formhals in 1934 and further he issued other patents on E-spin (US Patents 2,160,962 and 2,187,306) in 1939 and 1940 [35]. The spinning process invented by Formhals includes a moving thread collection device to collect the fibres in a stretched condition which is similar to the spinning drum in conventional E-spin [36]. In the past 50 years, about 60 patents have been filed on electrospinning solutions and melts and a simple equipment is invented which can generate highly electrified streams with uniform droplets of around 0.1 mm in diameter [37]. In 1955, Drozin studied the dispersion of different liquids in aerosols under high electrical field. After that, Simons invented a equipment in 1966 for the fabrication of ultra thin and extremely lightweight nonwoven fibres with different patterns under using electrical spinning and patented the work [38]. The "Taylor cone" model which theoretically modelled the conical geometry of liquid droplet, was established by Sir Geoffrey Ingram Taylor in 1966 [39]. In 1971, an apparatus was made by Baumgarten which fabricate electrospin acrylic fibre with diameters of 0.05 to 1.1 μm [40]. After that very small number of research articles and patents has been reported on E-spin technology in the 1980 and 1990s. The E-spin field has obtained significant attention and fast-growing development since Reneker et al. produced a number of fibres (diameter < 5 μm) by charging the polymer solution with high voltage [41]. As E-spin can easily generate fibres and fibrous structures with diameters in nano-range, it gained more attention may be due to the continuously increasing interest in nanotechnology [34]. All over the world, above 200 universities and research organizations are involved in studying different aspects of E-spin process and its applications. Also, some companies including Freudenberg and Donaldson Company have been utilizing E-spin process in the air filtration products for last two decades [42]. After 2010, major research progress on E-spin is focusing on the application of electrospun fibres on different fields using nanoparticle incorporated fibres, functionalized polymers, metal oxide composited nanofibres. A comprehensive overview of the advancement in E-spin technology is summarized in Table 1.

Table 1 Summary of historical development of E-spin technology

Year	Developments	References
1902	E-spin technique patented by Cooley and Morton	[43, 44]
1914	John Zeleny discovered jet ejection from the tip of the metal capillary	[31]
1934	Formhals patented the design of E-spin instrument	[32]
1936	Norton patented the air-blast fibre formation from melt in place of solution	[45]
1938–1940	Another three patents by Formhals on E-spin	[46–48]
1950–1959	Factory production of E-spin nanofibres used as filters for gas mask application	[49]
1960	Jet formation study of E-spin fibres for filtration material	[50]
1964	Taylor cone generation	[51]
1971	Baumgarten developed the apparatus to spin acrylic microfibres	[40]
1995	Diameter of fibres reduced with increasing distance from collector to needle tip and Taylor cone	[52]
1996–2001	Discovery related to the working parameters of E-spin like instrumental, solution and ambient parameters	[41, 53, 54]
2001–2005	From different countries, various E-spin instruments were developed by various researchers. Variety of synthesis and characterization techniques of E-spin nanofibres	[21, 37, 55, 56]
2006–2014	During this time, most of the researches have been focused on the application of E-spun nanofibres on energy, sensing, tissue engineering, catalyst and drug delivery	[55, 57–65]

3 Fundamentals of Electrospinning

The fundamental of E-spin is based on "**electrostatic interactions**". In detail, the high voltage power supply is connected to the peristaltic pump which pushes the solution from the needle. A counter electrode called collector collected single droplet ejected from the tip of the needle. It is obvious that the electrostatic interaction results in the fibre formation. Beside the polymeric nanofibres and composites, ceramic nanofibres can also be fabricated by changing the parameters including solution property, operational conditions and ambient property. The nanofibres fabricated by this technique can be associated with various chemical compositions, tuneable diameter and different morphology [55].

3.1 Working Principle

E-spin is a unique technique which uses electrostatic force of interaction to synthesize nanofibres from polymers and polymer solution or melts. The fibres are associated with smaller diameter and larger surface area than other conventional spinning techniques [66]. Different varieties of nanofibres including carbon materials, inorganic materials, polymers and their hybrid composites can be synthesized by this technique [67–69]. However, the basic theory of other spinning techniques like electrostatic precipitators and pesticide sprayers are similar to that of E-spin process which is based on electrostatic repulsive forces.

Generally, in E-spin process, an electrostatic voltage (around 5–30 kV) is engaged to charge the solution to start the emersion of liquid jet from the tip of the needle using a spinneret, such as coaxial, single and multi-spinneret [70, 71]. The polymer solution has its own surface tension into the syringe, and at the tip of the needle, it can be charged outside by applying high voltage power supply. The electrically conducting spinneret and oppositely charged collector are separated at an optimum distance (~10–25 cm) before the ejection of polymer solution. The liquid jet elongated through spiral loops, as the diameter of the loops increases, the jet becomes lengthier and thinner. With the evaporation of the solvent, solidification of the polymer precursor occurred and gets collected on the target [71–73]. The polymer molecules remain highly oriented in the fibres due to the stretching of jets, which have been established by polarized Raman and polarized FT-IR spectroscopy [74, 75].

However, these nanofibres are classified into two categories namely aligned and random nanofibres. Both of them have excellent properties like tuneable surface morphology, high surface-to-volume ratio and high porosity with interconnected pores [49, 76, 77]. The E-spin technique in which simple plate collectors are used, generally synthesizes nonwoven or random nanofibres and a disk or cylinder with faster rotating speed are involved produces aligned nanofibres. E-spin nanofibres have different morphologies including hollow, porous, sea-island, core–shell and dense structures according to various applied parameters like applied voltage, distance

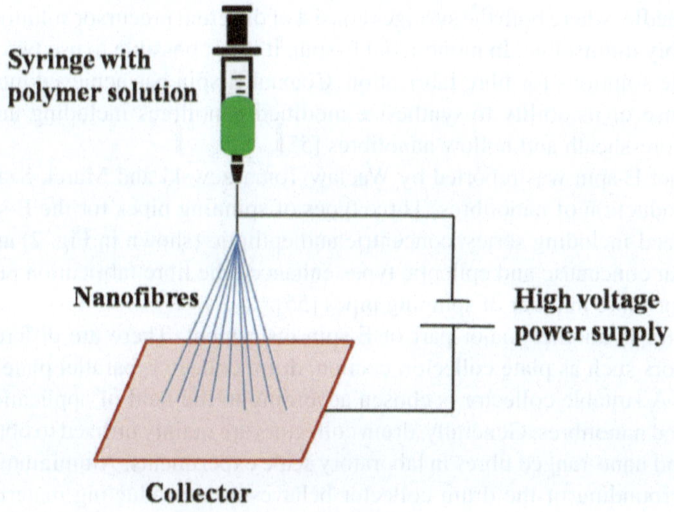

Fig. 1 Schematic representation of electrospinning process

between the collector and spinneret, feed rate and solution concentration [78]. Figure 1 illustrates the schematic representation of E-spin setup.

3.2 *Instrumental Setup*

The instrumentation of E-spin includes three main parts, first and the dominating part is high voltage power supply, second is the assembly of needle and syringe (known as spinneret in needleless type) and the last is collector. Now a days, various types of E-spin instruments are accessible in the market. They are different only in the design of spinneret and collector. Among them some include an electrode material as spinneret while most of the instruments are needleless spinneret.

However, two main categories of instruments are in practice such as horizontal and vertical. In case of horizontal type, the implied force is the charged force gained by applied potential and the attractive force of oppositely charged collector which draws the fibre. On the other hand, in vertical type, the fibres are pulled by two forces including collector charge and gravitational force which results thinner fibres with minimum diameter.

Another classification of E-spin instrument depends on number of nozzles like single nozzle E-spin and multi-nozzle E-spin. In single nozzle E-spin, only the solutions which are easily soluble could be used for fibre synthesis. Multi-nozzle is advantageous over the single nozzle due to the huge production of fibres. E-spin instruments can also be classified based on the number of axial units such as coaxial, mono-axial and multi-axial E-spin instrument. Coaxial E-spin employs two syringes and one needle, where both the syringes consist of different precursor solutions which are probably immiscible. In mono-axial E-spin, it is not possible to use two different immiscible solutions for fibre fabrication. Coaxial E-spin has achieved more attention because of its ability to synthesize modified nanofibres including uni-axially aligned, core-sheath and hollow nanofibres [55].

Multi-jet E-spin was reported by Waclaw Tomaszewski and Marek Szadkowski for the production of nanofibres. Three types of spinning pipes for the E-spin head were utilized including series, concentric and eplliptic (shown in Fig. 2) and it was proved that concentric and epileptic types enhanced the fibre fabrication process by using 10 or more number of spinning pipes [55].

Collector is another major part of E-spin instrument. There are different types of collectors such as plate collector, cocoon, drum collector, parallel plate and disc collector. A suitable collector is chosen according to the field of application of the synthesized nanofibres. Generally, drum collectors are mainly utilized to obtain well-aligned and nano-ranged fibres in laboratory scale experiments. Aluminium foil kept in the surrounding of the drum collector behaves as a conducting material which collects fibres [55].

According to the needle type of E-spin instrument, the instrumental parameters like solution volume, needle tip, modified syringe and gauge diameter could be selected. On the other hand, for large scale production of nanofibres, needleless E-spin is more beneficial in industries.

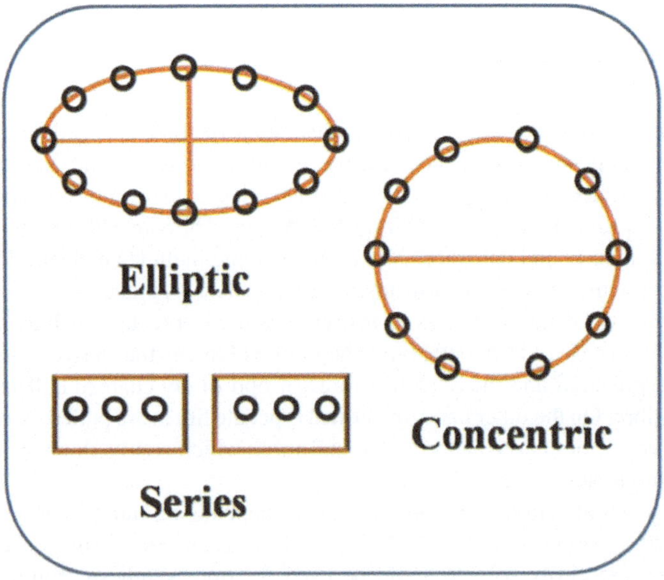

Fig. 2 Three types of spinning pipes for the E-spin head: elliptic, concentric and series. Redrawn from Ref. [55]

3.3 *Controlling Parameters*

Different parameters which influence the E-spin for fibre fabrication include instrumental parameters, solution parameters and ambient parameters. Category-wise factors influencing E-spin are summarized in Table 2.

3.3.1 Instrumental Parameters

Generally, lesser flow rate of the solution helps the polarization of the precursor solution. Flow rate is the flow of the solution under applied pressure. Increasing the flow rate of 1 ml/min results in the formation of beaded fibres. The morphology and diameter of fibres change with shape and diameter of the gauge respectively. Change in the gauge size affects the diameter of micro-scale fibres while change in gauge shape (elliptic and sphere) effects the morphology of fibres.

Change in applied potential effects the diameter of E-spin fibres. It has established by various research groups that with varying applied potential the diameter of fibre changes [79–81]. Influence of applied potential on the diameter of nanofibres is limited comparing with other instrumental parameters of E-spin. The distance between the tip of the needle and the collector allows enough opportunity for the evaporation of solvent. If the distance is lesser, thicker or beaded fibres can be formed while larger distance results in discontinuous fibres. So, by keeping other parameters constant, change in the distance from needle tip to the collector have a major influence on fibre morphology [82].

The variety of collector used including plate, rotating rods or wheels, pin, crossbar, disk, drum and liquid bath influences the orientation or alignment of fibres. Properly oriented nanofibres can be synthesized by utilizing a disk collector. A home-made collector was designed by Younan Xia et al. by applying a novel gap strategy for oriented fibres of polyvinyl pyrrolidone (PVP)/tetramethyl ammonium chloride involving two silicon strips (kept in certain distance) as collector [83]. In this method,

Table 2 List of influencing parameters in E-spin process

Instrumental parameters	Solution parameters	Ambient parameters
Flow rate (slower flow rate results smooth fibres)	Types of polymer (conducting and non-conducting, natural and synthetic)	Collector temperature
Gauge diameter	Molecular weight of polymer	Humidity
Applied potential (~5–30 kV)	Precursor solution and solvent	–
Distance between collector and tip of the needle	–	–
Variety of collector (plate, rotating rods or wheels, pin, crossbar, disk, drum, etc.)	Surface tension, conductivity and viscosity	–

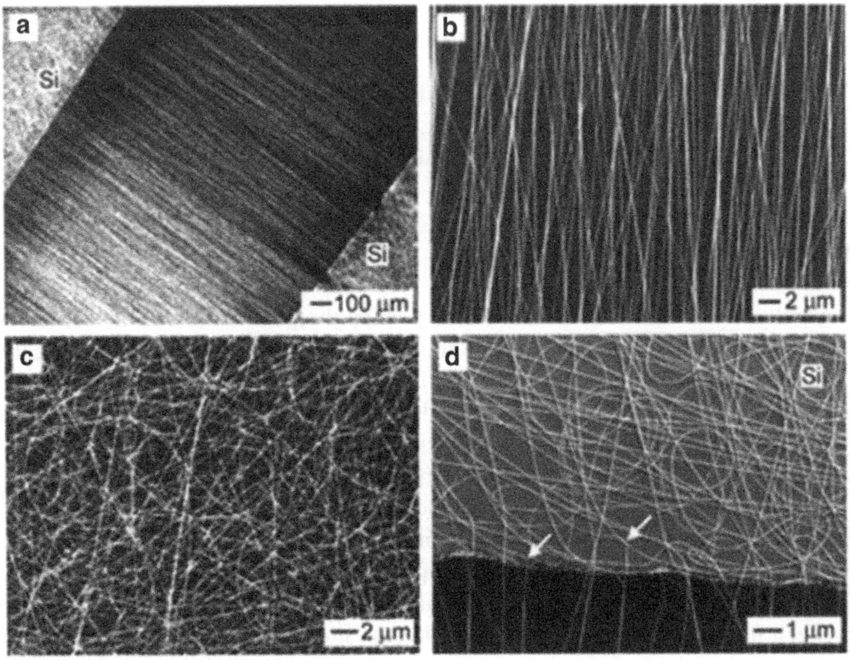

Fig. 3 **a** Dark-field optical micrograph of PVP NFs collected on top of a gap formed between two silicon stripes. **b, c** SEM images of same sample, showing NFs deposited (B) across the gap and (C) on top of the silicon stripe. **d** SEM image of NFs region close to the edge of the gap. Reprinted with permission from Ref. [83]

two types of forces are experienced by the nanofibres one from the charges on the surface of the silicon collector by the charged electrospun nanofibres and other from the splitting applied potential. They studied that the thicker nanofibres could be fabricated by reducing the gap between the silicon strips and increasing the collection time as shown in Fig. 3. It is interesting to observe that some of the nanofibres (marked by arrows in Fig. 3d) alter their direction perpendicularly to the edges before reaching to silicon collector. This variation in orientation further confirms the influence of two different forces on the nanofibres.

3.3.2 Solution Parameters

Depending on the application of nanofibres, suitable polymers are selected for E-spin. Nanotube enclosed nanofibres made up with conducting polymers obtain huge attention in the field of sensor, electrode materials and bio-active materials because of their capability trigger electrical charges remain inside the molecules. Kumber et al. fabricated nanofibres of chitosan, a crystalline, natural biopolymer [65]. They used derivative of 2-nitrobenzyl chitosan prepared by mixing 2-nitrobenzaldehyde with different concentration and chitosan for the fabrication of chitosan nanofibres. As synthesized imino-chitosan derivatives are very promising candidate for the shielding against yeast, bacteria and fungi.

The molecular weight of a selected polymer is an important factor which affects the properties of nanofibres. Same polymer having different molecular weight can synthesize nanofibres with different diameter [84, 85]. To obtain smooth and continuous nanofibres, the selected polymer should have optimized molecular weight. Sahoo and Panda had synthesized and characterized barium titanate nanofibre using E-spin technology [84, 85]. For a concentration dependant study, they used PVP (8–12% weight) with barium titanate composite. In Fig. 4, we can observe that with increasing the concentration of solution from 8 to 12%, the fibre diameter increased by 50 nm because of the viscosity change of the polymer solution.

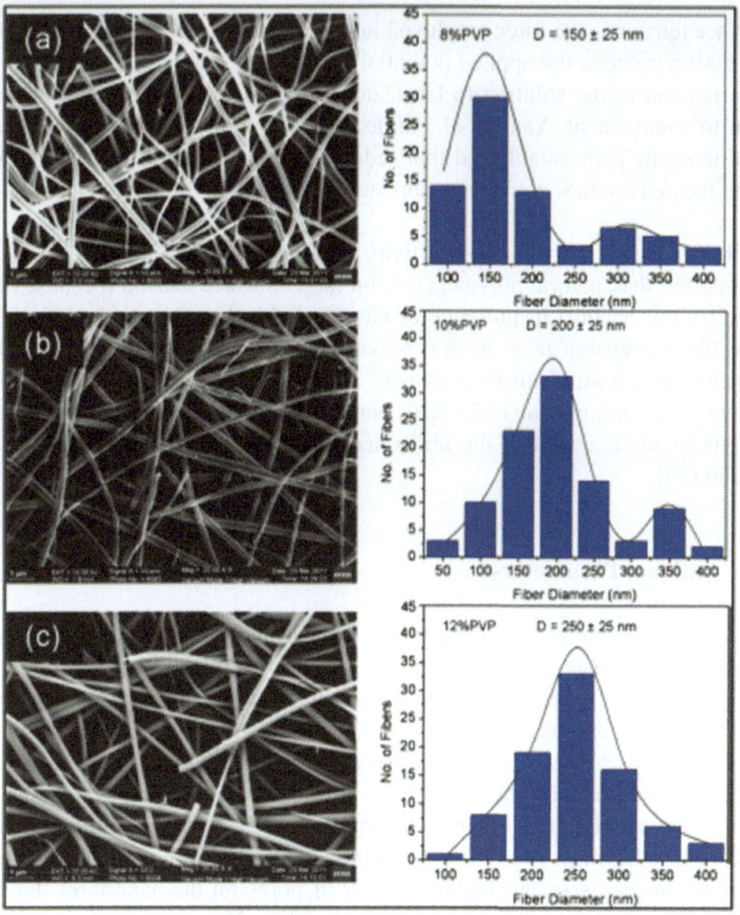

Fig. 4 SEM images and their corresponding diameter distributions of PVP–$BaTiO_3$ nanofibres with **a** 8%, **b** 10% and **c** 12% PVP. Reprinted with permission from Ref. [84]

For the synthesis of composite nanofibres, precursor solution is mixed with the polymer solution and delivered to E-spin for synthesis. The precursor can be nanoparticles or metal salt solution. They are blended with required solvent or polymer solution to obtain suitable viscosity for E-spin. During the travelling from the tip of the needle to the collector, the solvent evaporated which results in the fabrication of nanofibres. Due to this reason, specific solvent required to be selected for a certain precursor.

The main correlated solution properties that directly affect the E-spin ability of the solution are surface tension, conductivity and viscosity. One of the delicate features of fibre preparation technology which can be controlled by changing the concentration of the polymer solution is the viscosity of the spinning solution [86]. Fibres with micro- and nanoscale diameters can be obtained only with solutions of optimum viscosity. On the other hand, very high or very low viscosity solutions result beaded fibres.

Surface tension is the force employed in the plane of surface per unit length [87]. In the E-spin process, the applied potential should by high enough to overcome the surface tension of the solution to fabricate fibres. Solvents also affect the surface tension to some extent. Yang et al. studied that different solvent employ different surface tension. They established that reducing surface tension of the solution by keeping the concentration constant can convert the beaded fibres into smooth fibres [88].

In fibre preparation process conductivity of the solution also plays a key role. Wang et al. reported that, natural polymers exhibit larger surface tension (because of their polyelectrolytic nature) in presence of electric field, than the synthetic polymers. So, the fibres prepared from natural polymers are of poor quality compared with that synthesized from synthetic polymers. Conductivity of a spinning solution can be enhanced by mixing salts like KBr, NaCl, and KCl. Generally, with increasing conductivity, the diameter of the fibres decreases that means thinner fibres can be produced [89].

3.3.3 Ambient Parameters

Influence of collector temperature on the fibre properties is an important parameter in E-spin. Kim et al. established the impact of collector temperature on the porosity of E-spin fibres [90]. They studied that the porous nature of nonwoven nanofibres synthesized from poly-L-lactic acid (PLLA) in methylene chloride solvent was prominently affected by the temperature of the collector. The pores were developed in the nanofibres by the evaporation of the solvent molecules remain on the surface of the nanofibres, while the temperature reached to the boiling point of the solvent used (shown in Fig. 5a). With further increase in temperature, there was an increase in pore size and higher number of pores on the nanofibres due to the enhanced evaporation of the solvent molecules (Fig. 5b, c). When the temperature reached to 60 °C, there was a slight decrease in pore size due to the evaporation of solvent molecules present within the viscose polymer by the increment of the

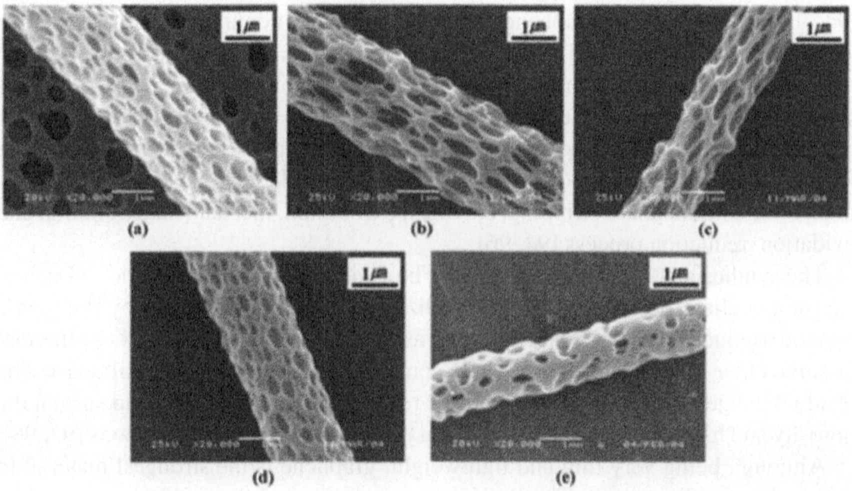

Fig. 5 SEM images of E-spun PLLA fibres as a function of collector temperature; a room temperature (21 °C), b 40 °C, c 50 °C, d 60 °C and e 70 °C. Reprinted with permission from Ref. [90]

volatility of the solvent (shown in Fig. 5d). When the temperature of the collector increases to the glass transition temperature (Tg) of the polymer, the porous structure of the fibres disorganized extremely due to the inadequate solidification of PLLA and high mobility (Fig. 5e).

In fibre fabrication technique humidity and temperature involve as interconnected ambient parameters. Enhancement in temperature will cause decrease in humidity and helps to evaporate the solvent quicker. Higher humidity results in thicker fibres with larger diameter. In another report, Casper et al. studied that with increasing humidity porous nanofibres can be produced. The optimum humidity for the production of polystyrene nanofibres was less than 25% [91]. Therefore, to obtain appropriate porous nanofibres, optimum humidity is required to maintain.

4 What is Graphene?

World's first 2D, lightweight, paper-like material is known as "graphene" [92]. Various multidirectional properties including conductivity, high surface area, mechanical strength, hydrophobicity, specific capacitance, photocatalytic and antibacterial activity makes graphene far more potential candidate than other nanomaterials [93]. Graphene is one atom thick, hexagonal layer of sp^2 hybridized carbon atoms. The carbon atoms are densely packed and looks like a honeycomb lattice.

A stable single-layered graphene was first prepared by mechanical stripping in 2004. For this discovery, Geim and Novoselov won Noble prize in physics in 2010

[94, 95]. Single-layered graphene is lattice iss purely conjugated with alternative single and double bond and the π electron clouds remain delocalized over the sp^2 hybridized carbon atoms. Graphene is the thinnest nanomaterial in the world with a stable single layer having thickness only 0.334 nm [94]. However, it is very difficult to prepare monolayered, pure graphene, some techniques are reported in the literature for the synthesis of single or few layered graphene like chemical vapour deposition (CVD), liquid phase stripping, chemical exfoliation, mechanical exfoliation and oxidation–reduction process [94, 96].

The synthesis of graphene can roughly be divided into three categories. Graphite was first oxidized by strong acids and oxidizing agents to fabricate GO by Hammer's method. Reduced graphene oxide (rGO) was synthesized by reducing GO by thermal or some other reduction process, where some oxygen functionalities still exist. The residual oxygen functionalities inhibit the re-stacking of rGO sheets and sustain the porosity and high surface area, which is very crucial for energy applications [97, 98].

Although being very thin and lightweight, graphene is the strongest material in the world ever discovered. The intrinsic strength of graphene is 42 Nm^{-1} and it is 200 times stronger than steel [99]. It is so robust that it can overshadow the hardness of diamond. A single sheet of graphene is transparent but it is able to absorb light [100]. It is elastic and pliable with Young's modulus of around 1 TPa [99]. But it is impermeable to all the liquids and gases except water [101]. Graphene is also associated with excellent thermal conductivity in the range of ~ $(4.84 \pm 0.44) \times 10^3$ to $(5.30 \pm 0.48) \times 10^3$ Wm^{-1} K^{-1}[102]. These values refer that; graphene is far more superior to carbon nanotubes in thermal conductivity. Another crucial property of graphene is excellent electrical conductivity with high intrinsic mobility of 2×10^5 cm^2v^{-1} s^{-1}[103, 104].

There are several practical applications of graphene as a versatile material in science, technology and environmental aspect including electronics, photonics, energy storage and generation, sensing, biomedical, optoelectronics and flexible electronics [99]. Different properties and application of graphene in various field are shown schematically in Fig. 6.

4.1 Graphene as Nanofiller in Electrospinning

The incorporation of GO into the electrospun nanofibres was started around this decade. Graphene is a favourable candidate to perform as nanofiller in E-spin to achieve desired nanofibres due to its versatile properties like mechanical, sensing, biomedical and electrical. Graphene has been proved to be a potential nanofiller candidate due to its multifunctional properties, which strongly reorganize the lightweight nanocomposites. Various types of natural and synthetic polymers were electrospun utilizing graphene nanofiller which exceptionally enhance the E-spin technique and had exhibited extraordinary enhancement in the characteristics of nanofibres like hydrophilicity, mechanical strength, conductivity, mechanical strength and thermal stability [105, 106]. The packing of graphene in an E-spin

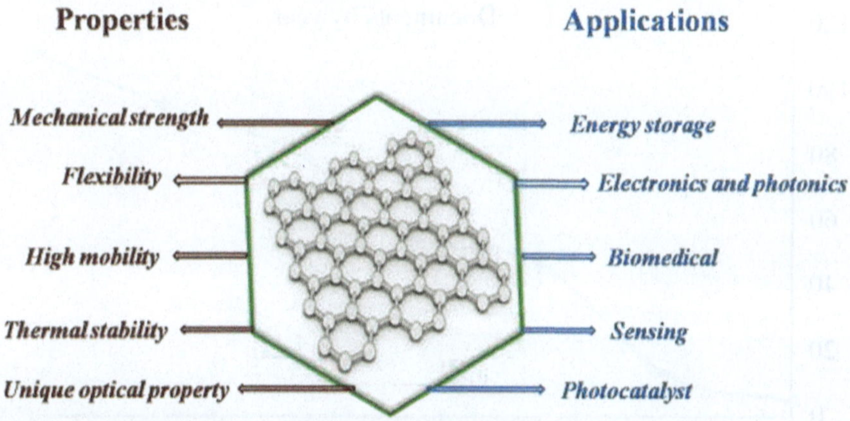

Fig. 6 Properties and applications of graphene

process is an important step which decides that chemical affinity, flexibility, stability and functionality. This involves two steps, first is GO sheets can introduce into the polymer solution by solution blending, in-situ polymerization or melt mixing and the second is synthesized nanofibres were reduced by annealing under high temperature or by chemical method which is mentioned as rGO nanofibres. Actually GO is a bed electrical conductor and an insulator. But when it is treated with strong reducing agent or annealed at high temperate, most of the conjugated structure of graphene has been re-established by the elimination of oxygen functionalities. However, there are several methods reported where reduced graphene has been directly used with polymer solution for E-spin [107, 108]. But, this process faces the drawback of inhomogeneous dispersion which may leads to difficulties and challenges for uninterrupted E-spin process. An overview of graphene used as nanofiller reported in the literature from 2010 to 2016 is shown in Fig. 7 [109].

4.2 Graphene-Based Electrospun Nanofibres

There are extensive advancements in graphene-based electrospun conductive nanofibres, especially in the area of electronics. Comparing with conventional metallic wires graphene-based conducting nanofibres are famous material because of their amazing properties including high electrical property, mechanical strength, lightweight and environmental stability. For the fabrication of graphene-based conductive, flexible nanofibres, the precursor solution was prepared using polymers including polyvinyl alcohol (PVA), polyacrylonitrile (PAN), polyvinyl chloride (PVC), poly(vinyl acetate) (PVAc), poly(lactic-co-glycolic acid) (PLGA) and poly(methyl methacrylate) (PMMA) which are summarized in Table 3 with the

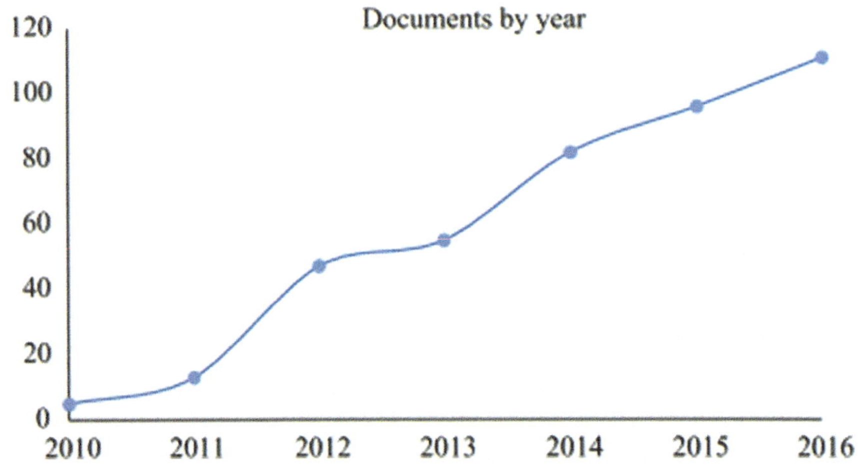

Fig. 7 Number of articles published by year using graphene as a nanofiller data analysis carried out using the Scopus search system with the term "graphene electrospinning" from 2010 to 2016. Reused with permission from Ref. [109]

specific reduction method. It is well established that while graphene is combined with fibres, it enhances their mechanical and electrical properties.

Matsumoto reported a methodical interaction of PAN with graphene [115]. By the opening of multiwalled carbon nanotubes (MWCNTs) using oxidation process, they prepared graphene oxide nanoribbons (GONRs) and then electrospun with GONRs in PAN/dimethylformamide (DMF) solution. Due to the thermal reduction, the electrical conductivity of graphene was hugely enhanced with highest obtained conductivity of 165.10 S cm^{-1}. The reduction strategy and sheet to sheet interdependence into the fibres effectively enhance the conductivity of fibres. The main challenges to achieve graphene-based nanofibres are the enhancement in dispersion, proper loading of GO and alignment within the polymer matrix. Along with the thermal reduction, many attempts were carried out on chemical reduction of graphene-based electrospun nanofibres. A chemical reduction of graphen-base nanofibres was reported by Wang et al., where they recovered the valuable conductive network of graphene [111]. A composite framework of GO sheets with polyvinyl alcohol (PVA) and PAN was fabricated using hydrazine hydride (N_2H_4) as reducing agent. The SEM images of GO and graphene composited PVA and PAN electrospun nanofibres are shown in Fig. 8. Applying this chemical reduction method, the problem of dispersion of graphene can be eliminated as well as enhanced conductivities can be obtained. For example, graphene-PAN composite offered conductivity of 75 S cm^{-1} and graphene-PVA offered 25 S cm^{-1}.

The major operating parameter to influence the characteristics of graphene-based nanofibres is the interaction between graphene and polymer matrix. This can be tuned by using selective chemical treatment on the surface of the nanofiller [111]. It is observed that the conductivity of nanofibres significantly enhance with high

Table 3 Summary of graphene-based nanofibre composites and their reduction methods

Electrospun nanofibre composites	Reduction method	References
PAN and GONRs	Annealed at 1000 °C for 1 h	[110]
GO, PAN and PVP	N_2H_4 for 6 h	[111]
PANi/G-PBASE and PMMA	Hydrazine monohydrate at 60 °C for 24 h	[112]
S-RGO/ACF	Heated at 800 °C for 1 h	[113]
PI-GNR/CNT	HI.H_2O at 98 °C for 10 h	[114]
PVC/PLGA nanofibres	HI solution (55%) at 100 °C for 1 h	[115]
RuO_2/ACNF and G	Heated at 800 °C for 1 h	[116]
GCNF @MoS_2	Heated at 800 °C for 2 h	[117]
RGO/PAN	Heated at 800 °C for 1 h	[107]
PAN/Fe_2O_3/G	Carbonized at 650 °C for 1 h	[118]
PAN/PMMA, $SbCl_3$ and GO	Heated at 700 °C for 2 h	[119]
CNF, Si and graphene-covered Ni particles	Carbonized at 650 °C for 1 h	[120]
PANi and PAN with G and GO nanosheets	Ammonia solution at 180 °C for 1 h	[121]
GO polyamide 66	0.1 wt % N_2H_4 and annealing at 350 °C	[122]
GNSs and AgNPs	$NaBH_4$ at 100 °C for 24 h	[123]
PANi with HCSA and PEO filled G-PBASE	N_2H_4 heated at 80 °C for 24 h	[124]

*PAN-Polyacrylonitrile; GONRs-Graphene oxide nanoribbons; GO-Graphene oxide; PVP-Polyvinyl pyrrolidone; PANi-Polyaniline; G-PBASE-Graphene with 1-pyrene-butanoic acid succinimidyl ester; PMMA-Poly(methyl methacrylate); S-RGO-Ultrasonic spray reduced graphene oxide; ACF-Anisotropic conductive film; PI-GNR-Polyamide-graphene nanoribbons; CNT-Carbon nanotube; PVC-Polyvinyl chloride; PLGA-Poly(lactic-co-glycolic acid); ACNF-Activated carbon nanofibre; G-Graphene; GCNF-Graphene-wrapped electrospun carbon nanofibres; RGO-Reduced graphene oxide; CNF-Carbon nanofibre; GNSs-Graphene nanosheets; AgNPs-Silver nanoparticles; HCSA-Camphor-10-sulphonic acid; PEO-Poly(ethylene oxide)

temperature reduction comparing with chemical method of reduction. The reason may be the reduction of huge number of oxygen containing functional groups from the GO surface, which remain attached into the interior of the aromatic zone of GO and recovering the sp^2 configuration of graphene which satisfactorily enhance the conductivity of nanofibres by the thermal reduction methods.

For further improvement of conductivity, metal ions can be composited with graphene-based electrospun nanofibres. This can be synthesized by mixing different metals including Si, Ag, Ni, Sn and Ru with the precursor solutions. In most of the cases, graphene-based electrospun nanofibres are used as back-bone materials for the introduction of electroactive materials like conducting polymers, metal oxides including MnO_2, RuO_2, Fe_2O_3 and Co_3O_4 obtain ultrahigh conductivity [125–127]. Moayeri and Ajji reported that coaxial E-spin is an alternative and beneficial method

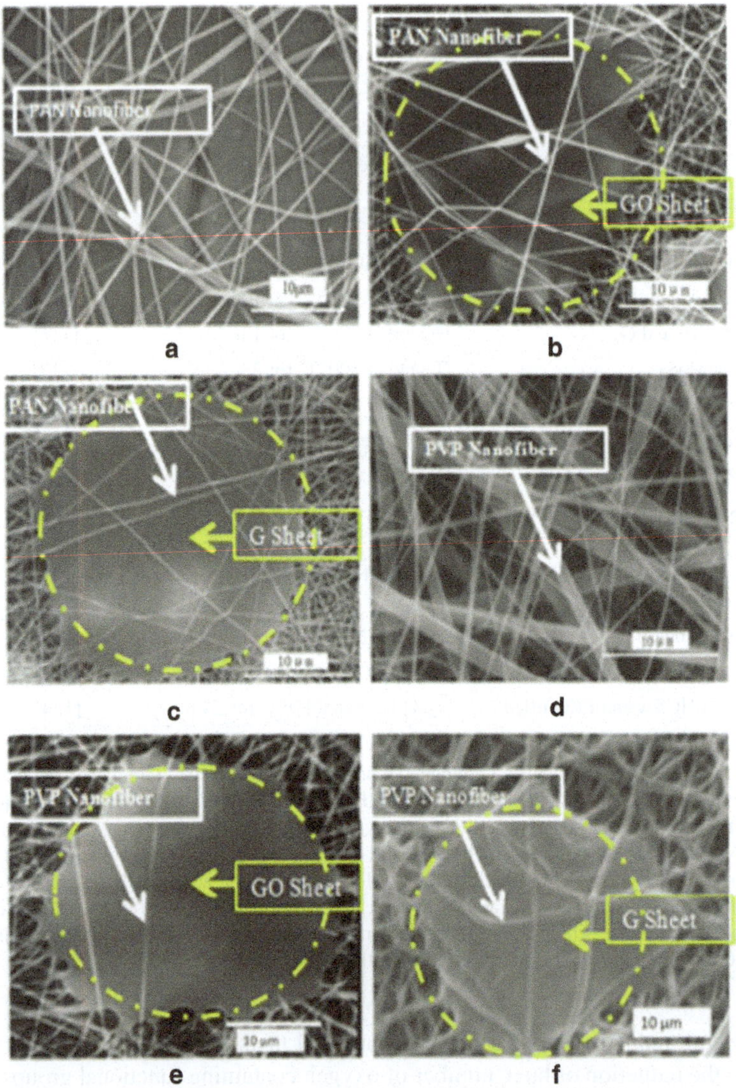

Fig. 8 SEM images of **a** PAN electrospun nanofibres, **b** GO-PAN electrospun nanofibre composite, **c** graphene-PAN nanofibre composite, **d** PVP electrospun nanofibres, **e** GO-PVP electrospun nanofibre composite, and **f** graphene-PVP nanofibre composite adapted from Ref. [111]

of preparing graphene composited conductive polymer polyaniline (PANi) with novel core–shell structure, compared with unidirectional E-spin [112]. They used reduced graphene with 1-pyrene-butanoic acid succinimidyl ester (PBASE) to synthesize nanofibres named as PANi/G-PBASE and the conductivity of nanofibres had boosted up to 30 S cm^{-1}.

Synthesis of GO is also a very crucial factor which influences the conductivity of nanofibres. This inspires the scientists to investigate new methods to functionalize nanofibres with GO to enhance the conductivity of electrospun nanofibres. For the industrial production of nanofibres where large scale production is required and there is a need to use huge amount of GO. In this case, researchers use a mist of GO applying an ultrasonic atomizer to incorporate graphene into the nanofibres because of relatively easy synthesis process for fabrication of large scale product with desired quality. Wang et al. followed an impressing strategy where they synthesize nanofibres simultaneously using E-spin and ultrasonication to dope GO using an ultrasonic atomizer by spraying (shown in Fig. 9) and the achieved conductivity was up to 0.42 S cm^{-1}[113].

On the other hand, oriented graphene sheets and anisotropic materials inside the nanofibres have an important role on the conductivity like contrastive conductivity of hierarchical graphitic GO in various directions. Due to this property, network-like structures are constructed from cross-linked graphene in perpendicular as well as parallel directions because of the unidirectional emission of electrons. This phenomena was addressed by Liu et al. who fabricated carbon nanotubes (CNTs) and polyamide-graphene nanoribbons (PI-GNR) to study the anisotropic behaviour of GO [114]. They demonstrated electrical conductivities in two different directions; in perpendicular direction 7.2×10^{-8} S cm^{-1} and in parallel direction 8.3×10^{-2} S cm^{-1}.

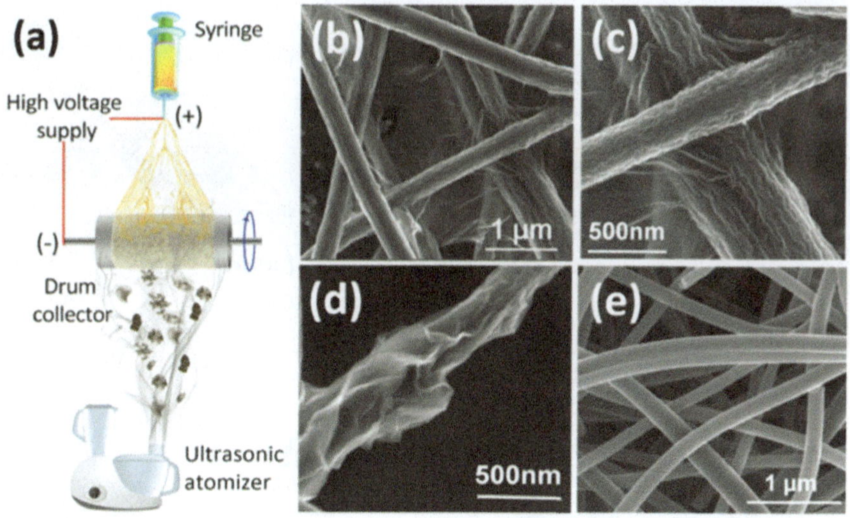

Fig. 9 **a** Schematic of the electrospinning technique coupled with ultrasonic spraying; **b** top-view SEM image of the as-made S-RGO/ACF; **c**, **d** SEM images of the individual S-RGO/ACF fibre; **e** pristine ACF surface in the absence of graphene. Reprinted with permission from Ref. [113]

4.3 Application in Various Fields

Graphene-based electrospun nanofibres are morphologically modified by required treatments according to the selective application. The aspect of E-spin using graphene-based nanofibres is able to contribute in various fields of applications including chemical, biomedical, defence and environmental applications.

4.3.1 Biomedical Application

Tissue engineering is very essential for improvement and repairing of injured tissues in human body. The wide application of E-spin technique in tissue engineering is no longer new. GO/polymer composites also attract much attention in this field due to having properties like noncytotoxicity and biocompatibility. In this field, scaffolds play a crucial role as they allow support for cell fitting and also permit cell growing into the tissue until it is capable to support itself. Different types of GO/polymer composites have been fabricated as scaffolds for tissue engineering including PAN/rGO [128], PVA/GO [129], poly (lactic acid) (PLA)/polyurethane (PU)/GO [105] and PVA/chitosan (CS)/GO [130]. Schematic of the synthesis procedure of PVA/chitosan (CS)/GO is shown in Fig. 10. Due to the presence of GO, polymeric scaffolds have excellent mechanical, electrical and thermal properties. Cell

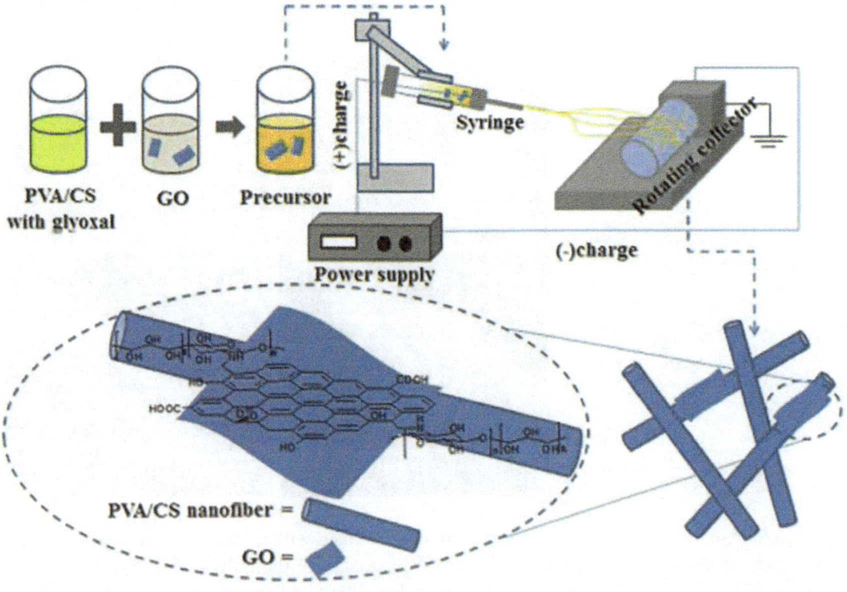

Fig. 10 Schematic representation of the synthesis procedure of PVA/CS nanofibres with self-assembled GO. Reproduced with permission from Ref. [130]

adhesion and proliferation on these GO/polymer scaffolds were studied morphologically and quantitatively. Cells including adipose-derived stem cells, osteoblastic cells and skeletal myoblasts cells were cultured on the samples. They reported that a little amount of GO did not limit the viability and proliferation of cells which described the sufficient cell affinity of GO. Cells are expected to spread on the scaffolds and some of them improve the cell growth by attaining the same cell proliferation rate to that of tissue culture plates. In recent years, scaffolds for skin tissues and musculoskeletal have attracted the attention of researchers with aims to support the patients with injuries. Tissues or organs without recovering ability are a vast area of tissue engineering which needs to be more explored. One of its branches is vascular tissue engineering which aims to synthesize functional vascular grafts used in vivo to replace blood vessels and support them to regenerate. Currently available commercial synthetic grafts used by surgeons are extended polytetrafluoroethylene (ePTFE) and woven poly (ethylene terephthalate) (Dacron). Although the motto was to replace blood vessels with large diameter, they were not suitable for grafts with small diameter because of the intimal hyperplasia and thrombosis risks. Jing et al. synthesized thermoplastic polyurethane (TPU)/GO scaffold utilizing grounded rotation mandrel as E-spin collector which resulted tubular scaffolds with small diameter [131]. With increasing the amount of GO, hydrophilicity of scaffolds and tensile strength increased.

Another efficient function of scaffolds which required to be enlarged is the capability of controlled drug release onto selective tissues and organs. For efficient transportation of drugs in a targeted drug delivery system, scaffolds can reduce systemic toxicity and improve efficiency of the drugs. E-spin nanofibres are very efficient materials having higher drug encapsulation capability and higher stability, because of their high porosity and high surface area. For anticancer drug delivery, an electrospun scaffold was studied utilizing polyethylene oxide (PEO)/CS/GO nanocomposite [132]. In this carrier system, GO played the role of nanocarrier by monitoring the controlled release of doxorubicin (DOX), an anthracycline antibiotic applied in chemotherapy for intravenous administration. GO surface makes a strong bond with DOX by $\pi - \pi$ stacking interactions, which allow the controlled release of drug. GO also contains functional groups like carboxylic and epoxy groups which can interact with the amine groups of CS. They allowed high drug loading of 98% because of the $\pi - \pi$ stacking interactions among GO and DOX with the pores in nanofibrous scaffolds. A faster drug release at pH 5.3, because of the instability of hydrogen bonding between GO and DOX.

On the other hand, **synthesis of scaffolds** is also succeeded to provide a imitated environment to the original body system. In 2015, Luo et al. synthesized GO doped poly (lactic-co-glycolic acid) (PLGA) nanofibre scaffold added with human marrow mesenchymal stem cells (MSCs) for oestrogenic differentiation [133]. The diameter and the porous structure of the synthesized fibres are similar to the morphological structure with the original extracellular matrices (ECM) and enhance cell proliferation and attachment. GO takes part in two different roles, first is improving protein inducer adsorption ability and hydrophilic capability of nanofibres and second is enhancing the adhesion, proliferation and differentiation of human MSCs towards osteoblasts.

4.3.2 Chemical Application

Graphene-based nanofibres attain a huge interest in biosensing application due to its novel properties like excellent electrocatalytic activity, mechanical strength, high surface area and easy to functionalize. In 2015, Ding et al. fabricated a sensor for Cu (II) detector made up with rGO/polyvinyl butyral (PVB) nanofibres [134]. Electrospun nanofibres of GO/PVB on glassy carbon electrode (GCE) were reduced electrochemically to fabricate rGO/PVB. One of the heavy metals Cu (II) coming from machinery, mining and metal smelting, contaminate water sources. Though Cu (II) is an essential element in human body, larger intake of Cu (II) may cause Menke's syndrome and Wilson's disease [134]. The prepared sensor made up with rGO/PVB nanofibres exhibited good analytical activity with a lower detection limit of 4.1 nM (S/N = 3), linear range of 0.06–22 μM, selectivity and reproducibility RSD = 0.49% and large sensitivity of 103.51 μA. μM^{-1} cm^{-2}. Zhang et al. reported a nanofibrous membrane of graphene quantum dots (GQD) by E-spin of the solution of GQD mixed with PVA [135]. For the first time the fabricated a PVA/GQD nanofibre membrane utilized as dual-purpose electrochemical and fluorescent biosensors for highly sensitive detection of glucose and hydrogen peroxide (H_2O_2). They reported that the prepared biosensors showed high performance with lower detection limit of 10.0 and 1.0 μM and linear detection ranges of 0.25–24 and 0.05–35 mM for glucose and H_2O_2 biosensors respectively.

Energy storage is one of the most essential fields which meet the need of in our modern society. The constantly increasing demand of rechargeable and sustainable energy sources to power up variety of energy storage systems. Thangappan et al. reported a electrospun hybrid material using GO, PVP and vanadium acetylacetonate [136]. The synthesized core–shell graphene oxide/vanadium pentoxide (GVO) nanofibrous mats used as electrode material in three-electrode cell and results an excellent specific capacitance of 453.824 F g^{-1} (shown in Fig. 11). Another core–shell nanofibrous structure was synthesized which contains ZnO/carrier polymer (core) and rGO/PAN (shell) [137]. The presence of ZnO enhances the theoretical capacity at 978 mA h g^{-1}. But it is also associated with large volume expansion around 228% which results destruction of mechanical integrity. Taking the same solution, they encapsulated ZnO nanoparticles inside the glassy carbon-reduced graphene oxide (C-rGO) hollow core. The void spaces allow a buffer zone which adjust the volume change due to lithiation/delithiation of ZnO and stop friction of nanoparticles by fragmentation and detachment. The free standing electrode exhibited capacitance of 815 mA h g^{-1} at 50 mA g^{-1} current density and the capacity retention was around 80% after 100 cycles. Another study showed that the combination of MnO_2 with graphene become very relevant as electrode material for supercapacitor application [138]. However, they offered poor intrinsic conductivity of 10^{-5} – 10^{-6} S cm^{-1}. They designed electrospun nanofibre of MnO_2 and hierarchical porous

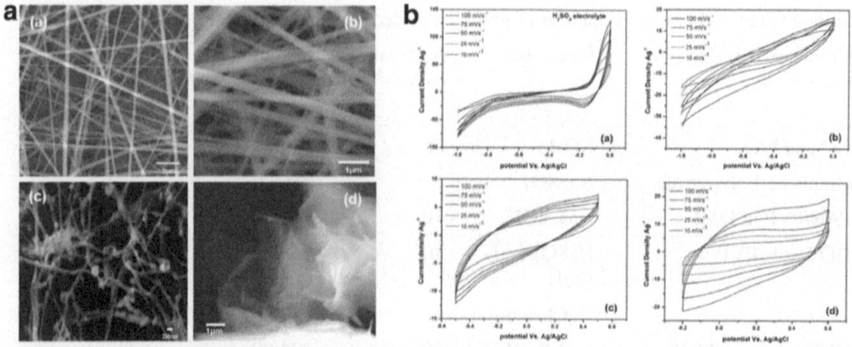

Fig. 11 **a** SEM images of (a) pure G/VO nanofibres, (b) annealed nanofibres at 350 °C and (c) 550 °C, (d) SEM of pure GO and **b** CV curves of (a, b) graphene and V_2O_5 nanofibres in different electrolytes (KOH and H_2SO_4) and CV curves for (c, d) pure V_2O_5 nanofibres and GO. Reused with permission from Ref. [136]

carbon nanofibre (CNF)/graphene which showed specific capacitance of 210 F g^{-1} at 1 mA cm^{-2} current density. Mixing of graphene with conducting polymer like polypyrrole (PPy) deposited on the surface of a CNF composite through an easy electrodeposition process can be a suitable material for pseudocapacitors. Gan et al. reported this type of core–shell structure which was synthesized in two steps [138]. At first CNF was synthesized by E-spin using PAN/dimethyl formamide (DMF) solution and then coating of nanofibres on graphene/PPy through electrodeposition. The electrode material showed specific capacitance of 386 F g^{-1}. Electrochemical performances of hybrid graphene-based E-spun nanofibres are enlisted in Table 4.

Shape memory nanofibre combined with GO was synthesized with shape memory PU (SMPU) by Tan et al. [154]. They showed that comparing with SMPU nanofibrous mat, SMPU/GO exhibited superior shape memory effect and lesser thermal shrinkage. The nanofibre composite exhibited thermal shrinkage as low as $4.7 \pm 0.3\%$ when GO loading reached to 4 wt %, while the recovery and average fixation ratio were 96.5 and 92.1% respectively. They reported GO as a efficient candidate with enhanced property as nanofiller for shape memory nanofibres.

A corrosion protection coating using composite of nylon-6, 6 with functionalised GO was reported by Campos et al. [155]. The electrospun nylon-6, 6 with 2% GO loading exhibited coating capacitance of 10^{-7} F cm^{-2}, while the same for coating prepared by deposition was 10^{-4} F cm^{-2}. These values suggest that the charge storage capacity of deposited coating is better than E-spin coating. However, main drawbacks of PU for its application are low mechanical property and low hydrophilicity. In such cases, GO-based polymer nanocomposites as nanofillers show effective improvement in this field. E-spin synthesis of in-situ PU/GO was coated on the surface of metallic stents by Pant et al. [106]. They showed that introduction of little amount of GO sheets into PU nanofibres could increase the stability of coating on the surface of the metallic stent. So, the optimum incorporation of GO sheets into PU can enhance the hydrophilicity without any toxic biological effect.

Table 4 Electrochemical properties of graphene-based E-spun hybrid nanofibres

Graphene-based hybrid nanofibres	Electrolyte	Scan rate (mVs^{-1})	Specific capacitance ($F g^{-1}$)	References
GO and PAN	KOH	25	146.62	[139]
Graphene-Polypyrrole	Na_2SO_4	2	386.00	[140]
PAN/PVP in DMF	6 M KOH	10	265.00	[141]
GO/V_2O_5/PVP	H_2SO_4 and KOH	10	453.82	[136]
NG-CNF	NaCl solution	1	337.85	[142]
CNF/graphene/MnO_2	6 M KOH	50	225.00	[138]
M-rGO/PA66	H_2SO_4	10	280.00	[143]
G/CNF-PAN	KOH	100	263.70	[144]
PVA/GO-PEDOT	1 M aq KCl	5	224.27	[145]
GO-PANi-PVDF	H_2SO_4	10	170.63	[146]
GO/PAN fibre paper	KOH	5	241.00	[147]
Graphene/RuO_2/ACNF composite	6 M KOH	25	180.00	[115]
MnO_2/HPCNF/G	KOH	10	210.00	[148]
rGO-thorn like TiO_2 nanofibre	1 M Na_2SO_4	5	178.00	[149]
GO-CNTs embedded in PAN	0.5 M Na_2SO_4	100	120.00	[150]
Graphene-TEOS-PAN	KOH	25	144.79	[151]
GNW-CNT-PAN	H_2SO_4	10	176.00	[152]
Graphene/PAN/PMMA	KOH	25	128.00	[153]

GO—Graphene oxide; PAN—Polyacrylonitrile; PVP—Polyvinyl pyrrolidone; DMF—Dimethyl formamide; NG—Nitrogen doped reduced graphene oxide; CNF—Carbon nanofibre, rGO—Reduced graphene oxide; PA66—Polyamide 66; G—Graphene; PVA—Polyvinyl alcohol; PEDOT—Poly(3,4-ethylenedioxythiophene); PAN—Polyaniline; PVDF—Polyvinylidene difluoride; ACNF—Activated carbon nanofibre; HPCNF—Hierarchical porous carbon nanofibre; CNTs—Carbon nanotubes; TEOS—Tetraethyl orthosilicate; GNW—Graphene-like carbon nanowall; PMMA—Polymethyl methacrylate.

Graphene base electropun nanofibres have been emerged as efficient candidate for photocatalysis. Zinc oxide decorated graphene nanofibres (G-ZnO) were first time synthesied by E-spin An et al. [156]. The fabricated composite mat showed excellent photocatalytic activity with 0.5 wt % graphene loading through degradation of methylene blue in presence of UV light. The best result was obtained with 0.5 wt % G-ZnO annealed at 400 °C and 80% degradation of methylene blue was obtained after 4 h of UV irradiation. Zhang et al. reported the photocatalyst using 1D TiO_2/GO electrospun nanocomposite and PVP polymer solution [157]. Continuous E-spin of TiO_2 nanofibres segregated by 5 wt % dispersed GO solution while PVP was removed by annealing at 500 °C. From photoluminescent study, lower excitation intensity for TiO_2/GO than only TiO_2 showed that recombination of photoinduced holes and electrons in TiO_2 could be efficiently limited in TiO_2/GO. Under visible light, the nanofibre composite exhibited better photocatalytic activity and greater mobility of charge carriers than bare TiO_2. With increasing GO concentration in nanofibres composite enhanced the photocatalytic activity of TiO_2/GO.

4.3.3 Defence Application

Protective clothing is a family of textile structure which guards the wearer from nuclear, chemical, physical and pathogenic hazards. Some of the most foremost types of protective clothing are fire fighters' protective clothing (used to protect against flame and heat), cold circumstances clothing, Ballistic protective clothing (used to protect against projectiles), sports clothing like scuba diving, swimming and NBC protection suits (used to protect against chemical, nuclear, biological warfare agents) [158]. Fan et al. synthesized aramid nanofibres by dissolving bulk aramid fibres in dimethylsulfoxide (DMSO) and potassium hydroxide (KOH) [159]. Aramid nanofibres-functionalized graphene sheets (ANFGS) was successfully fabricated as an efficient nanofiller for polymer reinforcement. The Young's modulus and the tensile strength of ANFGS/PMMA composite film reached 3.42 GPa and 63.2 MPa which were increased by 70.6% and 84.5% respectively, with 0.7 wt % loading of PMMA. It was also reported that, with increasing ANFGS loading, thermal stability of ANFGS/PMMA nanofibre composite increases. Due to the ability of ultraviolet light absorption of aramid nanfibres, ANFGS/PMMA composite have a degree of ultraviolet light shielding.

Nanofibre-based pH sensor was fabricated by E-spin method using PU/poly-2-acrylamido-2-methylpropanesulphonic acid (PAMPS)/(GO) with indicator dye [160]. Response time of sensing reduced drastically with increasing the loading concentration of GO and PAMPS. The hybrid nanofibrous PU/PAMPS/GO membrane are able to response instantly with the pH change of the solution.

4.3.4 Environmental Application

Desalination of water through air gap membrane using graphene-loaded nanofibre membrane was reported by Woo et al. [161]. To obtain superhydrophobic membrane, the concentration of GO loading into polyvinylidene fluoride-co-hexafluoropropylene electrospun membrane was optimized. With 5 wt % GO loading, the nanocomposite membrane results a stable and high air gap membrane distillation flux of 22.9 L m^{-2} h^{-1} and impressive salt rejection of 100% for 60 h of function with NaCl solution (3.5 wt %) as feed.

Wang et al. fabricated a nanofiltration membrane for **water purification** by the incorporation of graphene oxide sheets onto the surface of PAN electrospun nanofibres through vacuum suction method [162]. The designed membrane showed water flux of 8.2 L m^2h^{-1} bar^{-1} and had a contact angle of 61°. Due to the presence of both hydrophobic gates and hydrophilic nanochannels of GO, the diffusion of water through the composite membrane enhances. The membrane exhibited high rejection towards salts like NaCl and Na_2SO_4, and organic dye Congo red.

5 Future Scope

E-spin has emerged as an efficient technology enabling the scientists to know more about the characteristics of the nanofibre materials. Future advances in E-spin are mainly monitored by applications, which need special nanofibre morphology, chemistry and their scale up to industrial production. Different new innovations for E-spin are being developed to improve the applicability of these fibres. Some of these innovations include core shelled E-spin, mixing and multiple E-spin, blow assisted E-spin and coaxial E-spin. In coaxial, coaxial spinneret is used to synthesize nanofibres from two polymers and as a result shell of one polymer and core of another polymer is generated. This process achieves more attention because it generates combined polymers in radial and axial directions with novel characteristics. However, graphene-based electrospun nanofibres are one of the most dominant types of nanocomposites for their unique properties. Though variety of applications are already reported including biomedical, photocatalytic and sensing field, more attention is needed in catalysts in organic reactions and renewable energy storage devices, fine chemicals and pharmaceuticals. It is expected that graphene-based electrospun nanofibres could be used as catalyst in organic synthesis reactions and that would make the multistep synthesis process of pharmaceuticals easier.

The requirements for developing high-performance electrospun nanofibre composites are proper dispersion of GO and appropriate reduction of GO-based nanofibres. An appropriate dispersion of GO in the polymer matrix is a very complicated process. The dispersion of GO with a specific loading through mechanical mixing or sonication required to explore new solvent system with better contact. Another challenge is the reduction of GO-based nanofibres which is definitely a vital step to obtain a better quality nanofibre. Different methods have been reported and they have some advantages and disadvantages. Optimized result can be obtained through annealing treatment but proper reduction cannot achieve by chemical method which needs to be more explored by further efforts.

6 Conclusion

E-spin is a very simple, cost-effective and versatile technique that results in nonwoven nanofibres with high surface area, large pore volume and tuneable porosity. Solution and the influencing parameters like applied voltage, molecular weight, concentration of polymer and distance between tip to collector significantly affect the fibre characteristics and by optimizing these parameters, desired fibres can be fabricated. Graphene as a most promising material among all the allotropes of carbon has emerged as a efficient nanofiller for E-spin due to having excellent multidirectional properties like mechanical strength, electrical conductivity, high surface area and thermal stability. In this chapter, we describe a typical overview of advancement of E-spin technology for graphene-based materials and their application in different fields including biomedical, chemical, environmental and defence applications. Although having several advantages of E-spin, there are some limitations like lack of appropriate cellular infiltration in the fibres and small pore size. In general, considering the advantages of E-spin technology to fabricated graphene-based nanofibrous composite can be attributed as a new technique for next generation applications in laboratories and industries.

References

1. Hunley, M.T., Long, T.E.: Electrospinning functional nanoscale fibers: a perspective for the future. Polym. Int. **57**, 385–389 (2008)
2. Reneker, D.H., Yarin, A.L.: Electrospinning jets and polymer nanofibers. Polymer **49**, 2387–2425 (2008)
3. Persano, L., Camposeo, A., Pisignano, D.: Advancing the science and technology of electrospinning and functional nanofibers. Macromol. Mater. Eng. **302**, 1700237 (2017)
4. He, J.-H., Wan, Y.-Q., Yu, J.-Y.: Scaling law in electrospinning: relationship between electric current and solution flow rate. Polymer **46**, 2799–2801 (2005)
5. Zussman, E., Theron, A., Yarin, A.: Formation of nanofiber crossbars in electrospinning. Appl. Phys. Lett. **82**, 973–975 (2003)
6. Theron, S., Yarin, A., Zussman, E., Kroll, E.: Multiple jets in electrospinning: experiment and modeling. Polymer **46**, 2889–2899 (2005)
7. Teo, W.E., Ramakrishna, S.: A review on electrospinning design and nanofibre assemblies. Nanotechnology **17**, R89 (2006)
8. Welle, A., Kröger, M., Döring, M., Niederer, K., Pindel, E., Chronakis, I.S.: Electrospun aliphatic polycarbonates as tailored tissue scaffold materials. Biomaterials **28**, 2211–2219 (2007)
9. Wu, Y., He, J.-H., Xu, L., Yu, J.-Y.: Electrospinning drug-loaded poly (Butylenes Succinate-cobytylene Terephthalate)(PBST) with acetylsalicylic acid (aspirin). Int. J. Electrospun Nanofibers Appl. **1**, 1–6 (2007)
10. Barnes, C., Sell, S., Knapp, D., Walpoth, B., Brand, D., Bowlin, G.: Preliminary investigation of electrospun collagen and polydioxanone for vascular tissue engineering applications. Int. J. Electrospun Nanofibers Appl. **1**, 73–87 (2007)
11. Cui, W., Zhou, S., Li, X., Weng, J.: Drug-loaded biodegradable polymeric nanofibers prepared by electrospinning, p. 1070. Tissue Engineering, Mary Ann Liebert, Inc 140 Huguenot Street, 3RD Fl, New Rochelle, NY 10801 USA (2006)
12. Liang, D., Hsiao, B.S., Chu, B.: Functional electrospun nanofibrous scaffolds for biomedical applications. Adv. Drug Deliv. Rev. **59**, 1392–1412 (2007)
13. Chong, E.J., Phan, T.T., Lim, I.J., Zhang, Y., Bay, B.H., Ramakrishna, S., et al.: Evaluation of electrospun PCL/gelatin nanofibrous scaffold for wound healing and layered dermal reconstitution. Acta Biomater. **3**, 321–330 (2007)
14. Ma, Z., Kotaki, M., Ramakrishna, S.: Electrospun cellulose nanofiber as affinity membrane. J. Membr. Sci. **265**, 115–123 (2005)

15. Matthews, J.A., Wnek, G.E., Simpson, D.G., Bowlin, G.L.: Electrospinning of collagen nanofibers. Biomacromol **3**, 232–238 (2002)
16. Stankus, J.J., Guan, J., Wagner, W.R.: Fabrication of biodegradable elastomeric scaffolds with sub-micron morphologies. J. Biomed. Mater. Res. Part A Official J. Soc. Biomater. Jpn. Soc. Biomater. Aus. Soc. Biomater. Korean Soc. Biomater. **70**, 603–614 (2004)
17. Chen, Z., Mo, X., Qing, F.: Electrospinning of collagen–chitosan complex. Mater. Lett. **61**, 3490–3494 (2007)
18. Alessandrino, A., Marelli, B., Arosio, C., Fare, S., Tanzi, M.C., Freddi, G.: Electrospun silk fibroin mats for tissue engineering. Eng. Life Sci. **8**, 219–225 (2008)
19. Yang, F., Murugan, R., Wang, S., Ramakrishna, S.: Electrospinning of nano/micro scale poly (L-lactic acid) aligned fibers and their potential in neural tissue engineering. Biomaterials **26**, 2603–2610 (2005)
20. Liu, Y., He, J.-H.: Bubble electrospinning for mass production of nanofibers. J. Nano Res. **23**, 125–128 (2007)
21. Tomaszewski, W., Szadkowski, M.: Investigation of electrospinning with the use of a multi-jet electrospinning head. Fibres Text. Eastern Europe **13**, 22 (2005)
22. Varabhas, J., Chase, G.G., Reneker, D.: Electrospun nanofibers from a porous hollow tube. Polymer **49**, 4226–4229 (2008)
23. Dosunmu, O., Chase, G.G., Kataphinan, W., Reneker, D.: Electrospinning of polymer nanofibres from multiple jets on a porous tubular surface. Nanotechnology **17**, 1123 (2006)
24. Eichhorn, S.J., Sampson, W.W.: Statistical geometry of pores and statistics of porous nanofibrous assemblies. J. R. Soc. Interface **2**, 309–318 (2005)
25. Ekaputra, A.K., Prestwich, G.D., Cool, S.M., Hutmacher, D.W.: Combining electrospun scaffolds with electrosprayed hydrogels leads to three-dimensional cellularization of hybrid constructs. Biomacromol **9**, 2097–2103 (2008)
26. Barzegar, F., Bello, A., Fabiane, M., Khamlich, S., Momodu, D., Taghizadeh, F., et al.: Preparation and characterization of poly (vinyl alcohol)/graphene nanofibers synthesized by electrospinning. J. Phys. Chem. Solids **77**, 139–145 (2015)
27. Wahab, I.F., Razak, S., Azmi, N.S., Dahli, F.N., Yusof, A.H.M., Nayan, N.H.M.: Electrospun graphene oxide-based nanofibres. Adv. Carbon Nanostruct. **10** (2016)
28. Zhang, H., Yu, X., Guo, D., Qu, B., Zhang, M., Li, Q., et al.: Synthesis of bacteria promoted reduced graphene oxide-nickel sulfide networks for advanced supercapacitors. ACS Appl. Mater. Interfaces **5**, 7335–7340 (2013)
29. Navarro-Pardo, F., Martinez-Hernandez, A.L., Velasco-Santos, C.: Carbon nanotube and graphene based polyamide electrospun nanocomposites: a review. J. Nanomater. **2016** (2016)
30. Bhardwaj, N., Kundu, S.C.: Electrospinning: a fascinating fiber fabrication technique. Biotechnol. Adv. **28**, 325–347 (2010)
31. Zeleny, J.: The electrical discharge from liquid points, and a hydrostatic method of measuring the electric intensity at their surfaces. Phys. Rev. **3**, 69 (1914)
32. Anton, F.: Process and apparatus for preparing artificial threads. Google Patents (1934)
33. Taylor, G.I.: Electrically driven jets. Proc. Royal Soc. London Math. Phys. Sci. **313**, 453–475 (1969)
34. Huang, Z.-M., Zhang, Y.-Z., Kotaki, M., Ramakrishna, S.: A review on polymer nanofibers by electrospinning and their applications in nanocomposites. Compos. Sci. Technol. **63**, 2223–2253 (2003)
35. Pawlowski, K.J., Barnes, C.P., Boland, E.D., Wnek, G.E., Bowlin, G.L.: Biomedical Nanoscience: Electrospinning Basic Concepts, Applications, and Classroom Demonstration, p. 827. MRS Online Proceedings Library Archive (2004)
36. Subbiah, T., Bhat, G.S., Tock, R.W., Parameswaran, S., Ramkumar, S.S.: Electrospinning of nanofibers. J. Appl. Polym. Sci. **96**, 557–569 (2005)
37. Li, D., Xia, Y.: Electrospinning of nanofibers: reinventing the wheel? Adv. Mater. **16**, 1151–1170 (2004)
38. Simons, H.L.: Process and apparatus for producing patterned non-woven fabrics. Google Patents (1966)

39. Taylor, G.I.: The force exerted by an electric field on a long cylindrical conductor. Proc. R. Soc. Lond. A **291**, 145–158 (1966)
40. Baumgarten, P.K.: Electrostatic spinning of acrylic microfibers. J. Colloid Interface Sci. **36**, 71–79 (1971)
41. Reneker, D.H., Chun, I.: Nanometre diameter fibres of polymer, produced by electrospinning. Nanotechnology **7**, 216 (1996)
42. Ramakrishna, S., Fujihara, K., Teo, W.-E., Yong, T., Ma, Z., Ramaseshan, R.: Electrospun nanofibers: solving global issues. Mater. Today **9**, 40–50 (2006)
43. Cooley, J.F.: Apparatus for electrically dispersing fluids. Google Patents (1902)
44. Morton, W.J.: Method of dispersing fluids. Google Patents (1902)
45. Norton, C.L.: Method of and apparatus for producing fibrous or filamentary material. Google Patents (1936)
46. Anton, F.: Method and apparatus for the production of fibers. Google Patents (1938)
47. Anton, F.: Method and apparatus for spinning. Google Patents (1939)
48. Anton, F.: Artificial thread and method of producing same. Google Patents (1940)
49. Luo, C., Stoyanov, S.D., Stride, E., Pelan, E., Edirisinghe, M.: Electrospinning versus fibre production methods: from specifics to technological convergence. Chem. Soc. Rev. **41**, 4708–4735 (2012)
50. Filatov, Y., Budyka, A., Kirichenko, V.: Electrospinning of micro-and nanofibers: fundamentals in separation and filtration processes. J. Eng. Fibers Fabrics. **3**, 488 (2007)
51. Taylor, G.I.: Disintegration of water drops in an electric field. Proc. R. Soc. Lond. A **280**, 383–397 (1964)
52. Doshi, J., Reneker, D.H.: Electrospinning process and applications of electrospun fibers. In: Conference Record of the 1993 IEEE Industry Applications Conference Twenty-Eighth IAS Annual Meeting. IEEE, pp. 1698–703 (1993)
53. Fong, H., Reneker, D.H.: Elastomeric nanofibers of styrene–butadiene–styrene triblock copolymer. J. Polym. Sci. Part B Polym. Phys. **37**, 3488–3493 (1999)
54. Bognitzki, M., Czado, W., Frese, T., Schaper, A., Hellwig, M., Steinhart, M., et al.: Nanostructured fibers via electrospinning. Adv. Mater. **13**, 70–72 (2001)
55. Thenmozhi, S., Dharmaraj, N., Kadirvelu, K., Kim, H.Y.: Electrospun nanofibers: new generation materials for advanced applications. Mater. Sci. Eng. B **217**, 36–48 (2017)
56. Zong, X., Kim, K., Fang, D., Ran, S., Hsiao, B.S., Chu, B.: Structure and process relationship of electrospun bioabsorbable nanofiber membranes. Polymer **43**, 4403–4412 (2002)
57. Kim, I.-D., Rothschild, A., Lee, B.H., Kim, D.Y., Jo, S.M., Tuller, H.L.: Ultrasensitive chemiresistors based on electrospun TiO_2 nanofibers. Nano Lett. **6**, 2009–2013 (2006)
58. Liao, S., Li, B., Ma, Z., Wei, H., Chan, C., Ramakrishna, S.: Biomimetic electrospun nanofibers for tissue regeneration. Biomed. Mater. **1**, R45 (2006)
59. Sundarrajan, S., Ramakrishna, S.: Fabrication of nanocomposite membranes from nanofibers and nanoparticles for protection against chemical warfare stimulants. J. Mater. Sci. **42**, 8400–8407 (2007)
60. Munir, M.M., Widiyandari, H., Iskandar, F., Okuyama, K.: Patterned indium tin oxide nanofiber films and their electrical and optical performance. Nanotechnology **19**, 375601 (2008)
61. Sill, T.J., Von Recum, H.A.: Electrospinning: applications in drug delivery and tissue engineering. Biomaterials **29**, 1989–2006 (2008)
62. Lin, D., Wu, H., Zhang, R., Pan, W.: Enhanced photocatalysis of electrospun Ag-ZnO heterostructured nanofibers. Chem. Mater. **21**, 3479–3484 (2009)
63. Park, J.-S.: Electrospinning and its applications. Adv. Nat. Sci. Nanosci. Nanotechnol. **1**, 043002 (2011)
64. Sahay, R., Kumar, P.S., Sridhar, R., Sundaramurthy, J., Venugopal, J., Mhaisalkar, S.G., et al.: Electrospun composite nanofibers and their multifaceted applications. J. Mater. Chem. **22**, 12953–12971 (2012)
65. Nada, A.A., James, R., Shelke, N.B., Harmon, M.D., Awad, H.M., Nagarale, R.K., et al.: A smart methodology to fabricate electrospun chitosan nanofiber matrices for regenerative engineering applications. Polym. Adv. Technol. **25**, 507–515 (2014)

66. Zhu, M., Xiong, R., Huang, C.: Bio-based and photocrosslinked electrospun antibacterial nanofibrous membranes for air filtration. Carbohyd. Polym. **205**, 55–62 (2019)
67. Zhou, S., Zhou, G., Jiang, S., Fan, P., Hou, H.: Flexible and refractory tantalum carbide-carbon electrospun nanofibers with high modulus and electric conductivity. Mater. Lett. **200**, 97–100 (2017)
68. Duan, G., Bagheri, A.R., Jiang, S., Golenser, J., Agarwal, S., Greiner, A.: Exploration of macroporous polymeric sponges as drug carriers. Biomacromol **18**, 3215–3221 (2017)
69. Liu, L., Bakhshi, H., Jiang, S., Schmalz, H., Agarwal, S.: Composite polymeric membranes with directionally embedded fibers for controlled dual actuation. Macromol. Rapid Commun. **39**, 1800082 (2018)
70. Ma, W., Zhao, J., Oderinde, O., Han, J., Liu, Z., Gao, B., et al.: Durable superhydrophobic and superoleophilic electrospun nanofibrous membrane for oil-water emulsion separation. J. Colloid Interface Sci. **532**, 12–23 (2018)
71. Zhu, M., Hua, D., Pan, H., Wang, F., Manshian, B., Soenen, S.J., et al.: Green electrospun and crosslinked poly (vinyl alcohol)/poly (acrylic acid) composite membranes for antibacterial effective air filtration. J. Colloid Interface Sci. **511**, 411–423 (2018)
72. Shi, X., Xu, Z., Huang, C., Wang, Y., Cui, Z.: Selective swelling of electrospun block copolymers: from perforated nanofibers to high flux and responsive ultrafiltration membranes. Macromolecules **51**, 2283–2292 (2018)
73. Jiang, Z., Zhang, H., Zhu, M., Lv, D., Yao, J., Xiong, R., et al.: Electrospun soy-protein-based nanofibrous membranes for effective antimicrobial air filtration. J. Appl. Polym. Sci. **135**, 45766 (2018)
74. Yang, H., Jiang, S., Fang, H., Hu, X., Duan, G., Hou, H.: Molecular orientation in aligned electrospun polyimide nanofibers by polarized FT-IR spectroscopy. Spectrochim. Acta Part A Mol. Biomol. Spectrosc. **200**, 339–344 (2018)
75. Xu, H., Jiang, S., Ding, C., Zhu, Y., Li, J., Hou, H.: High strength and high breaking load of single electrospun polyimide microfiber from water soluble precursor. Mater. Lett. **201**, 82–84 (2017)
76. Zhu, M., Han, J., Wang, F., Shao, W., Xiong, R., Zhang, Q., et al.: Electrospun nanofibers membranes for effective air filtration. Macromol. Mater. Eng. **302**, 1600353 (2017)
77. Ma, W., Samal, S.K., Liu, Z., Xiong, R., De Smedt, S.C., Bhushan, B., et al.: Dual pH-and ammonia-vapor-responsive electrospun nanofibrous membranes for oil-water separations. J. Membr. Sci. **537**, 128–139 (2017)
78. Yang, H., Kou, S.: Recent advances of flexible electrospun nanofibers-based electrodes for electrochemical supercapacitors: a minireview. Int. J. Electrochem. Sci. **14**, 7811–7831 (2019)
79. Buchko, C.J., Chen, L.C., Shen, Y., Martin, D.C.: Processing and microstructural characterization of porous biocompatible protein polymer thin films. Polymer **40**, 7397–7407 (1999)
80. Yuan, X., Zhang, Y., Dong, C., Sheng, J.: Morphology of ultrafine polysulfone fibers prepared by electrospinning. Polym. Int. **53**, 1704–1710 (2004)
81. Zhang, C., Yuan, X., Wu, L., Han, Y., Sheng, J.: Study on morphology of electrospun poly (vinyl alcohol) mats. Eur. Polym. J. **41**, 423–432 (2005)
82. Ki, C.S., Baek, D.H., Gang, K.D., Lee, K.H., Um, I.C., Park, Y.H.: Characterization of gelatin nanofiber prepared from gelatin–formic acid solution. Polymer **46**, 5094–5102 (2005)
83. Li, D., Wang, Y., Xia, Y.: Electrospinning of polymeric and ceramic nanofibers as uniaxially aligned arrays. Nano Lett. **3**, 1167–1171 (2003)
84. Sahoo, B., Panda, P.: Preparation and characterization of barium titanate nanofibers by electrospinning. Ceram. Int. **38**, 5189–5193 (2012)
85. Palangetic, L., Reddy, N.K., Srinivasan, S., Cohen, R.E., McKinley, G.H., Clasen, C.: Dispersity and spinnability: why highly polydisperse polymer solutions are desirable for electrospinning. Polymer **55**, 4920–4931 (2014)
86. Li, Z., Wang, C.: One-Dimensional Nanostructures: Electrospinning Technique and Unique Nanofibers. Springer (2013)
87. Ramakrishna, S.: An Introduction to Electrospinning and Nanofibers. World Scientific (2005)

88. Yang, Q., Li, Z., Hong, Y., Zhao, Y., Qiu, S., Wang, C., et al.: Influence of solvents on the formation of ultrathin uniform poly (vinyl pyrrolidone) nanofibers with electrospinning. J. Polym. Sci. Part B Polym. Phys. **42**, 3721–3726 (2004)
89. Wang, Z., Li, Z., Liu, L., Xu, X., Zhang, H., Wang, W., et al.: A novel alcohol detector based on ZrO_2-doped SnO_2 electrospun nanofibers. J. Am. Ceram. Soc. **93**, 634–637 (2010)
90. Kim, C.H., Jung, Y.H., Kim, H.Y., Lee, D.R., Dharmaraj, N., Choi, K.E.: Effect of collector temperature on the porous structure of electrospun fibers. Macromol. Res. **14**, 59–65 (2006)
91. Casper, C.L., Stephens, J.S., Tassi, N.G., Chase, D.B., Rabolt, J.F.: Controlling surface morphology of electrospun polystyrene fibers: effect of humidity and molecular weight in the electrospinning process. Macromolecules **37**, 573–578 (2004)
92. Novoselov, K.S., Fal, V., Colombo, L., Gellert, P., Schwab, M., Kim, K.: A roadmap for graphene. Nature **490**, 192–200 (2012)
93. Soldano, C., Mahmood, A., Dujardin, E.: Production, properties and potential of graphene. Carbon **48**, 2127–2150 (2010)
94. Ren, S., Rong, P., Yu, Q.: Preparations, properties and applications of graphene in functional devices: a concise review. Ceram. Int. **44**, 11940–11955 (2018)
95. Novoselov, K.S., Geim, A.K., Morozov, S.V., Jiang, D., Zhang, Y., Dubonos, S.V., et al.: Electric field effect in atomically thin carbon films. Science **306**, 666–9 (2004)
96. Amiri, A., Naraghi, M., Ahmadi, G., Soleymaniha, M., Shanbedi, M.: A review on liquid-phase exfoliation for scalable production of pure graphene, wrinkled, crumpled and functionalized graphene and challenges. FlatChem **8**, 40–71 (2018)
97. He, T., Meng, X., Nie, J., Tong, Y., Cai, K.: Thermally reduced graphene oxide electrochemically activated by bis-spiro quaternary alkyl ammonium for capacitors. ACS Appl. Mater. Interfaces **8**, 13865–13870 (2016)
98. Hantel, M.M., Nesper, R., Wokaun, A., Kötz, R.: In-situ XRD and dilatometry investigation of the formation of pillared graphene via electrochemical activation of partially reduced graphite oxide. Electrochim. Acta **134**, 459–470 (2014)
99. Lee, C., Wei, X., Kysar, J.W., Hone, J.: Measurement of the elastic properties and intrinsic strength of monolayer graphene. Science **321**, 385–8 (2008)
100. Nair, R.R., Blake, P., Grigorenko, A.N., Novoselov, K.S., Booth, T.J., Stauber, T., et al.: Fine structure constant defines visual transparency of graphene. Science **320**, 1308 (2008)
101. Bunch, J.S., Verbridge, S.S., Alden, J.S., Van Der Zande, A.M., Parpia, J.M., Craighead, H.G., et al.: Impermeable atomic membranes from graphene sheets. Nano Lett. **8**, 2458–2462 (2008)
102. Balandin, A.A., Ghosh, S., Bao, W., Calizo, I., Teweldebrhan, D., Miao, F., et al.: Superior thermal conductivity of single-layer graphene. Nano Lett. **8**, 902–907 (2008)
103. Chen, J.-H., Jang, C., Xiao, S., Ishigami, M., Fuhrer, M.S.: Intrinsic and extrinsic performance limits of graphene devices on SiO_2. Nat. Nanotechnol. **3**, 206–209 (2008)
104. Dürkop, T., Getty, S., Cobas, E., Fuhrer, M.: Extraordinary mobility in semiconducting carbon nanotubes. Nano Lett. **4**, 35–39 (2004)
105. An, X., Ma, H., Liu, B., Wang, J.: Graphene oxide reinforced polylactic acid/polyurethane antibacterial composites. J. Nanomater. **2013** (2013)
106. Pant, H.R., Pokharel, P., Joshi, M.K., Adhikari, S., Kim, H.J., Park, C.H., et al.: Processing and characterization of electrospun graphene oxide/polyurethane composite nanofibers for stent coating. Chem. Eng. J. **270**, 336–342 (2015)
107. Dong, Q., Wang, G., Qian, B., Hu, C., Wang, Y., Qiu, J.: Electrospun composites made of reduced graphene oxide and activated carbon nanofibers for capacitive deionization. Electrochim. Acta **137**, 388–394 (2014)
108. Lavanya, T., Satheesh, K., Dutta, M., Jaya, N.V., Fukata, N.: Superior photocatalytic performance of reduced graphene oxide wrapped electrospun anatase mesoporous TiO_2 nanofibers. J. Alloy. Compd. **615**, 643–650 (2014)
109. Javed, K., Oolo, M., Savest, N., Krumme, A.: A review on graphene-based electrospun conductive nanofibers, supercapacitors, anodes, and cathodes for lithium-ion batteries. Crit. Rev. Solid State Mater. Sci. **44**, 427–443 (2019)

110. Matsumoto, H., Imaizumi, S., Konosu, Y., Ashizawa, M., Minagawa, M., Tanioka, A., et al.: Electrospun composite nanofiber yarns containing oriented graphene nanoribbons. ACS Appl. Mater. Interfaces **5**, 6225–6231 (2013)
111. Wang, Y., Tang, J., Xie, S., Liu, J., Xin, Z., Liu, X., et al.: Leveling graphene sheets through electrospinning and their conductivity. RSC Adv. **5**, 42174–42177 (2015)
112. Moayeri, A., Ajji, A.: Core-shell structured graphene filled polyaniline/poly (methyl methacrylate) nanofibers by coaxial electrospinning. Nanosci. Nanotechnol. Lett. **8**, 129–134 (2016)
113. Wang, G., Dong, Q., Wu, T., Zhan, F., Zhou, M., Qiu, J.: Ultrasound-assisted preparation of electrospun carbon fiber/graphene electrodes for capacitive deionization: importance and unique role of electrical conductivity. Carbon **103**, 311–317 (2016)
114. Liu, M., Du, Y., Miao, Y.-E., Ding, Q., He, S., Tjiu, W.W., et al.: Anisotropic conductive films based on highly aligned polyimide fibers containing hybrid materials of graphene nanoribbons and carbon nanotubes. Nanoscale **7**, 1037–1046 (2015)
115. Jin, L., Wu, D., Kuddannaya, S., Zhang, Y., Wang, Z.: Fabrication, characterization, and biocompatibility of polymer cored reduced graphene oxide nanofibers. ACS Appl. Mater. Interfaces **8**, 5170–5177 (2016)
116. Yang, K.S., Kim, B.-H.: Highly conductive, porous RuO_2/activated carbon nanofiber composites containing graphene for electrochemical capacitor electrodes. Electrochim. Acta **186**, 337–344 (2015)
117. Gu, H., Huang, Y., Zuo, L., Fan, W., Liu, T.: Graphene sheets wrapped carbon nanofibers as a highly conductive three-dimensional framework for perpendicularly anchoring of MoS_2: advanced electrocatalysts for hydrogen evolution reaction. Electrochim. Acta **219**, 604–613 (2016)
118. Zhang, B., Xu, Z.-L., Kim, J.-K.: In situ grown graphitic carbon/Fe_2O_3/carbon nanofiber composites for high performance freestanding anodes in Li-ion batteries. RSC Adv. **4**, 12298–12301 (2014)
119. Tang, X., Yan, F., Wei, Y., Zhang, M., Wang, T., Zhang, T.: Encapsulating Sn_xSb nanoparticles in multichannel graphene-carbon fibers as flexible anodes to store lithium ions with high capacities. ACS Appl. Mater. Interfaces **7**, 21890–21897 (2015)
120. Xu, Z.-L., Zhang, B., Zhou, Z.-Q., Abouali, S., Garakani, M.A., Huang, J., et al.: Carbon nanofibers containing Si nanoparticles and graphene-covered Ni for high performance anodes in Li ion batteries. RSC Adv. **4**, 22359–22366 (2014)
121. Mahmoudifard, M., Soleimani, M., Hatamie, S., Zamanlui, S., Ranjbarvan, P., Vossoughi, M., et al.: The different fate of satellite cells on conductive composite electrospun nanofibers with graphene and graphene oxide nanosheets. Biomed. Mater. **11**, 025006 (2016)
122. Huang, Y.-L., Baji, A., Tien, H.-W., Yang, Y.-K., Yang, S.-Y., Ma, C.-C.M., et al.: Self-assembly of graphene onto electrospun polyamide 66 nanofibers as transparent conductive thin films. Nanotechnology **22**, 475603 (2011)
123. Huang, Y.-L., Baji, A., Tien, H.-W., Yang, Y.-K., Yang, S.-Y., Wu, S.-Y., et al.: Self-assembly of silver–graphene hybrid on electrospun polyurethane nanofibers as flexible transparent conductive thin films. Carbon **50**, 3473–3481 (2012)
124. Moayeri, A., Ajji, A.: Fabrication of polyaniline/poly (ethylene oxide)/non-covalently functionalized graphene nanofibers via electrospinning. Synth. Met. **200**, 7–15 (2015)
125. Bi, R.-R., Wu, X.-L., Cao, F.-F., Jiang, L.-Y., Guo, Y.-G., Wan, L.-J.: Highly dispersed RuO_2 nanoparticles on carbon nanotubes: facile synthesis and enhanced supercapacitance performance. J. Phys. Chem. C **114**, 2448–2451 (2010)
126. Shan, Y., Gao, L.: Formation and characterization of multi-walled carbon nanotubes/Co_3O_4 nanocomposites for supercapacitors. Mater. Chem. Phys. **103**, 206–210 (2007)
127. Yuan, L., Lu, X.-H., Xiao, X., Zhai, T., Dai, J., Zhang, F., et al.: Flexible solid-state supercapacitors based on carbon nanoparticles/MnO_2 nanorods hybrid structure. ACS Nano **6**, 656–661 (2012)
128. Jin, L., Yue, D., Xu, Z.-W., Liang, G., Zhang, Y., Zhang, J.-F., et al.: Fabrication, mechanical properties, and biocompatibility of reduced graphene oxide-reinforced nanofiber mats. RSC Adv. **4**, 35035–35041 (2014)

129. Qi, Y., Tai, Z., Sun, D., Chen, J., Ma, H., Yan, X., et al.: Fabrication and characterization of poly (vinyl alcohol)/graphene oxide nanofibrous biocomposite scaffolds. J. Appl. Polym. Sci. **127**, 1885–1894 (2013)
130. Liu, Y., Park, M., Shin, H.K., Pant, B., Choi, J., Park, Y.W., et al.: Facile preparation and characterization of poly (vinyl alcohol)/chitosan/graphene oxide biocomposite nanofibers. J. Ind. Eng. Chem. **20**, 4415–4420 (2014)
131. Jing, X., Mi, H.-Y., Salick, M.R., Cordie, T.M., Peng, X.-F., Turng, L.-S.: Electrospinning thermoplastic polyurethane/graphene oxide scaffolds for small diameter vascular graft applications. Mater. Sci. Eng. C **49**, 40–50 (2015)
132. Ardeshirzadeh, B., Anaraki, N.A., Irani, M., Rad, L.R., Shamshiri, S.: Controlled release of doxorubicin from electrospun PEO/chitosan/graphene oxide nanocomposite nanofibrous scaffolds. Mater. Sci. Eng. C **48**, 384–390 (2015)
133. Luo, Y., Shen, H., Fang, Y., Cao, Y., Huang, J., Zhang, M., et al.: Enhanced proliferation and osteogenic differentiation of mesenchymal stem cells on graphene oxide-incorporated electrospun poly (lactic-co-glycolic acid) nanofibrous mats. ACS Appl. Mater. Interfaces **7**, 6331–6339 (2015)
134. Ding, R., Luo, Z., Ma, X., Fan, X., Xue, L., Lin, X., et al.: High sensitive sensor fabricated by reduced graphene oxide/polyvinyl butyral nanofibers for detecting Cu (II) in water. Int. J. Anal. Chem. **2015** (2015)
135. Zhang, P., Zhao, X., Ji, Y., Ouyang, Z., Wen, X., Li, J., et al.: Electrospinning graphene quantum dots into a nanofibrous membrane for dual-purpose fluorescent and electrochemical biosensors. J. Mater. Chem. B **3**, 2487–2496 (2015)
136. Thangappan, R., Kalaiselvam, S., Elayaperumal, A., Jayavel, R.: Synthesis of graphene oxide/vanadium pentoxide composite nanofibers by electrospinning for supercapacitor applications. Solid State Ionics **268**, 321–325 (2014)
137. Basavaraja, B.M., Majumder, S.B., Sharma, A.: Electrospun hollow glassy carbon–reduced graphene oxide nanofibers with encapsulated ZnO nanoparticles: a free standing anode for Li-ion batteries. J. Mater. Chem. A **3**, 5344–5351 (2015)
138. Kim, Y.A., Kim, B.-H.: Capacitive properties of hierarchically structured carbon nanofiber/graphene/MnO_2 hybrid electrode with nitrogen and oxygen heteroatoms. Carbon **107**, 783–791 (2016)
139. Kim, B.-H., Yang, K.S.: Structure and electrochemical properties of electrospun carbon fiber composites containing graphene. J. Ind. Eng. Chem. **20**, 3474–3479 (2014)
140. Gan, J.K., Lim, Y.S., Pandikumar, A., Huang, N.M., Lim, H.N.: Graphene/polypyrrole-coated carbon nanofiber core–shell architecture electrode for electrochemical capacitors. RSC Adv. **5**, 12692–12699 (2015)
141. Moayeri, A., Ajji, A.: High capacitance carbon nanofibers from poly (acrylonitrile) and poly (vinylpyrrolidone)-functionalized graphene by electrospinning. J. Nanosci. Nanotechnol. **17**, 1820–1829 (2017)
142. Liu, Y., Xu, X., Lu, T., Sun, Z., Chua, D.H., Pan, L.: Nitrogen-doped electrospun reduced graphene oxide–carbon nanofiber composite for capacitive deionization. RSC Adv. **5**, 34117–34124 (2015)
143. Wang, Y.-S., Li, S.-M., Hsiao, S.-T., Liao, W.-H., Chen, P.-H., Yang, S.-Y., et al.: Integration of tailored reduced graphene oxide nanosheets and electrospun polyamide-66 nanofabrics for a flexible supercapacitor with high-volume-and high-area-specific capacitance. Carbon **73**, 87–98 (2014)
144. Zhou, Z., Wu, X.-F.: Graphene-beaded carbon nanofibers for use in supercapacitor electrodes: synthesis and electrochemical characterization. J. Power Sour. **222**, 410–416 (2013)
145. Abdah, M.A.A.M., Zubair, N.A., Azman, N.H.N., Sulaiman, Y.: Fabrication of PEDOT coated PVA-GO nanofiber for supercapacitor. Mater. Chem. Phys. **192**, 161–169 (2017)
146. Rose, A., Raghavan, N., Thangavel, S., Maheswari, B.U., Nair, D.P., Venugopal, G.: Investigation of cyclic voltammetry of graphene oxide/polyaniline/polyvinylidene fluoride nanofibers prepared via electrospinning. Mater. Sci. Semicond. Process. **31**, 281–286 (2015)

147. Xie, Q., Zhou, S., Wu, S., Zhang, Y., Zhao, P.: Supercapacitive behavior of laminar-structured carbon cloth with alternating graphene and hybrid nanofibers: a synergistic effect of graphene-coating and post-oxidization. Appl. Surf. Sci. **407**, 36–43 (2017)
148. Kim, B.-H.: MnO_2 decorated on electrospun carbon nanofiber/graphene composites as supercapacitor electrode materials. Synth. Met. **219**, 115–123 (2016)
149. Kim, T.-W., Park, S.-J.: Synthesis of reduced graphene oxide/thorn-like titanium dioxide nanofiber aerogels with enhanced electrochemical performance for supercapacitor. J. Colloid Interface Sci. **486**, 287–295 (2017)
150. Hsu, H.-C., Wang, C.-H., Chang, Y.-C., Hu, J.-H., Yao, B.-Y., Lin, C.-Y.: Graphene oxides and carbon nanotubes embedded in polyacrylonitrile-based carbon nanofibers used as electrodes for supercapacitor. J. Phys. Chem. Solids **85**, 62–68 (2015)
151. Kim, S.Y., Yang, K., Kim, B.-H.: Enhanced electrical capacitance of heteroatom-decorated nanoporous carbon nanofiber composites containing graphene. Electrochim. Acta **137**, 781–788 (2014)
152. Hsu, H.-C., Wang, C.-H., Nataraj, S., Huang, H.-C., Du, H.-Y., Chang, S.-T., et al.: Stand-up structure of graphene-like carbon nanowalls on CNT directly grown on polyacrylonitrile-based carbon fiber paper as supercapacitor. Diam. Relat. Mater. **25**, 176–179 (2012)
153. Kim, B.-H., Yang, K.S., Ferraris, J.P.: Highly conductive, mesoporous carbon nanofiber web as electrode material for high-performance supercapacitors. Electrochim. Acta **75**, 325–331 (2012)
154. Tan, L., Gan, L., Hu, J., Zhu, Y., Han, J.: Functional shape memory composite nanofibers with graphene oxide filler. Compos. A Appl. Sci. Manuf. **76**, 115–123 (2015)
155. Menchaca-Campos, C., García-Pérez, C., Castañeda, I., García-Sánchez, M.A., Guardián, R., Uruchurtu, J.: Nylon/graphene oxide electrospun composite coating. Int. J. Polym. Sci. **2013** (2013)
156. An, S., Joshi, B.N., Lee, M.W., Kim, N.Y., Yoon, S.S.: Electrospun graphene-ZnO nanofiber mats for photocatalysis applications. Appl. Surf. Sci. **294**, 24–28 (2014)
157. Zhang, L., Zhang, Q., Xie, H., Guo, J., Lyu, H., Li, Y., et al.: Electrospun titania nanofibers segregated by graphene oxide for improved visible light photocatalysis. Appl. Catal. B **201**, 470–478 (2017)
158. Gorji, M., Bagherzadeh, R., Fashandi, H.: Electrospun nanofibers in protective clothing. In: Electrospun Nanofibers, pp. 571–98. Elsevier
159. Fan, J., Shi, Z., Zhang, L., Wang, J., Yin, J.: Aramid nanofiber-functionalized graphene nanosheets for polymer reinforcement. Nanoscale **4**, 7046–7055 (2012)
160. Gorji, M., Sadeghianmaryan, A., Rajabinejad, H., Nasherolahkam, S., Chen, X.: Development of highly pH-sensitive hybrid membranes by simultaneous electrospinning of amphiphilic nanofibers reinforced with graphene oxide. J. Func. Biomater. **10**, 23 (2019)
161. Woo, Y.C., Tijing, L.D., Shim, W.-G., Choi, J.-S., Kim, S.-H., He, T., et al.: Water desalination using graphene-enhanced electrospun nanofiber membrane via air gap membrane distillation. J. Membr. Sci. **520**, 99–110 (2016)
162. Wang, J., Zhang, P., Liang, B., Liu, Y., Xu, T., Wang, L., et al.: Graphene oxide as an effective barrier on a porous nanofibrous membrane for water treatment. ACS Appl. Mater. Interfaces **8**, 6211–6218 (2016)

Fabrication and Application Prospective of Graphene Infused Polymeric Flexible, Stretchable and Transparent Devices

Prashant Rawat and Deju Zhu

Abstract Electrospinning technique is one of the most common methods adopted for the fabrication of ultrathin films with nano-sized fibres. Recently, graphene materials have attracted a great deal of research interest in different areas, especially in the field of device fabrication. This chapter deals with the electrospinning fabrication process of graphene-based hybrid films for the development of flexible, stretchable, as well as transparent devices. A brief overview of graphene materials, history of flexible and stretchable devices, and application of graphene infused devices are also discussed. Finally, the book chapter ends with an outlook, and future research scope in this field.

Keywords Graphene · Flexible devices · Sensors · Stretchable devices · Electrospinning

1 Introduction

The electronic circuits are the veins of any system or electrical-structure that helps to perform various functions. Over 50 years, these circuits have revolutionized entire technology and their applications in many sectors (such as optical electronics, bio-medical, defence applications etc.). Today, almost everything we are using is made of these circuits which are tremendously smarter, flexible/stretchable and most importantly smaller in size. Modern engineering and manufacturing technologies are playing a critical role in introducing electronic gadgets on a daily basis. However, advancement in any technology requires a precise combination of advanced materials and a multidisciplinary approach to maximize the efficiency of the device. In the same series, nano-scale structures or devises are expected to be the future materials with

P. Rawat (✉)
Key Laboratory for Green and Advanced Civil Engineering Materials and Application Technology of Hunan Province, College of Civil Engineering, Hunan University, Changsha 410082, P. R. China

D. Zhu
International Science Innovation Collaboration Base for Green and Advanced Civil Engineering Materials of Hunan Province, Hunan University, Changsha 410082, P. R. China

S. K. Tiwari et al. (eds.), *Electrospinning of Graphene*, Carbon Nanostructures,
https://doi.org/10.1007/978-3-030-75456-3_2

exceptional performance capabilities. Nanoparticles like carbon nanotubes, silicon and graphene are the most dominant nanomaterials in terms of future technology which are extraordinarily robust, flexible and highly conductive at the same time. Interestingly, graphene, which is a 2D sheet structure (in which carbon allotropes are arranged in hexagonal structure), can be a key material for smarter and flexible circuits. Nevertheless, manufacturing technique like 'electrospinning' has covered a long journey (reported in 1931), and after many advancements, this fabrication method is most researched process for developing substrates for flexible, ultrathin and transparent devices.

Therefore, it is interesting to investigate how useful these techniques are from the perspective of flexible and stretchable devices and their wide range application. Additionally, it is also essential to explore, if carbon-based nanostructures (especially graphene) are helpful to overcome the existing limitations (related to conductivity with flexibility) in modern time.

2 Flexible and Stretchable Devices: A Brief History

Development of flexible electronic devices started from the 1990s when researchers started developing organic and inorganic substrates over polymer films [1]. The earlier research focus was related to flexible devices and their use; mainly as organic materials (which are flexible by their fundamental nature) for the development of light emitted diodes (LED) [2, 3]. Gradually use of inorganic materials (by using solution-based or evaporation) on plastic substrates came in existence [4, 5]. Nevertheless, the earlier researches were more focused about measuring and analyzing the electrical, mechanical or fundamental characteristics of a newly born class of materials [6]. Interestingly, a similar research is going on with a new series of nanomaterials [7, 8] for better properties (electrical, thermal and mechanical) with more flexibility. Rogers et al. [9] in 2001 reported the first flexible paper-like display integrated with microencapsulated electrophoretic "inks". This study proved the possibility of printing electrical systems on paper (with high quality in a larger area) which started a new era of improved electronic processing devices. Further researchers have developed artificial e-skin in 2004 [10], efficient electrical devices (by dry-transfer printing method) [11], thin-film transistor-based pressure and temperature sensor (in 2005) [12] and heterogeneous 2D/3D electronic systems in 2006 [13].

With the development in nano-technology, carbon-based nanoparticles have proved their significance in terms of ultra-light weighted transparent materials with exceptional conductivity and strength. Kim et al. [14] used recently discovered 'graphene' sheets for fabricating transparent and stretchable electrodes with outstanding mechanical, electrical and optical properties. Cao et al. [15] used single-walled carbon nanotubes (synthesized by CVD process) in small to medium-scale integrated circuits. This investigation proved the significance and potential of nanomaterials in flexible integrated circuits. In similar research, Sekitani et al. [16, 17] have reported the use of carbon nanotubes for flexible and stretchable circuits. By the year 2010, application of these flexible and stretchable circuits has started in multiple areas like synthetic e-skin, bio-patches, solar-cell etc. Mannsfeld et al. [18] proposed a new type of pressure sensor device (synthetic skin) with polydimethylsiloxane (PDMS) films as a key element. Takei et al. [19] proposed a design (or nanowire platform) for flexible devices which may present a critical role in future manufacturing related nanowire printing technique. Kim et al. [20] suggested a new strategy for bio-interfaced systems based on ultra-thin electronic flexible films with conformal silk-supported PI electrode arrays. Yamada et al. [21] proposed a stretchable human motion-detection strain sensor based on carbon nanotubes film. This research projected equipment which is suitable for real-life applications and human-friendly as well. Flexible, stretchable and transparent conducting films infused with super aligned carbon nanotubes (SACNTs) proved the importance of nanomaterials for flexible and stretchable devices.

Han et al. [22] used graphene for manufacturing flexible organic light-emitting diodes (OLEDs) which proved a great potential of graphene-based anodes for high-performing flexible organic optoelectronics devices. In 2013, Chae et al. [23] combined graphene and single-walled carbon nanotube (with a wrinkled Al_2O_3 dielectric layer) for making transparent and stretchable transistors. This type of transistor sustained stretching and releasing up to 1000 time without losing its performance. Crumpled graphene-paper-based stretchable and high performance based supercapacitors [24] proved the significance of flexible capacitor devices over unconventional energy storage devices (ESDs) in 2014. Xu et al. [25] fabricated fiber-shaped zinc-air battery with aligned and cross-stacked CNTs for the first time, which provided exceptional electrochemical features. Lee et al. [26] have developed highly flexible and durable electrodes based on silver nanoparticles and graphene for mobile-optoelectronics. In addition, this research highlighted the complexity of manufacturing techniques which are high hinder for developing advanced flexible equipment. Hybrid nanoparticle graphene and graphene quantum dots (GQDs) based sensitive photodetectors [27] proved themselves as highly flexible and stretchable materials. This new strategy proved 2D inclusion of nanomaterials as a key technology for many applications like a solar cell, LEDs, transistors etc. Wang et al. [28] raised the issue of damage in supercapacitors and fabricated a self-healable graphene oxide infused supercapacitor. However, the fabrication of such devices is a challenge, but the method proposed in this investigation may be key to future technology.

In 2018, manufacturing of rechargeable solid-state zinc-air batteries emerged with flexible carbon cloth (derived from porous graphene skin) [29] proved the advancement in nano-technology fabrication techniques for electrocatalyst energy storing devices. Liang et al. [30] proposed a novel design (with high-strained, stretchable and self-healable capacities) of supercapacitor based on carbon nanotubes/graphene/polymerized polyaniline nanowires. The proposed supercapacitors sustained 180 and 80.2% tensile strain and capacitance retention even after ten healing cycles. Stretchable electroluminescent display embedded with graphene-based hybrid electrodes [31] proved the significance of highly stretchable electrodes for future technology. Snapp et al. [32] also focused on hybrid systems (infused

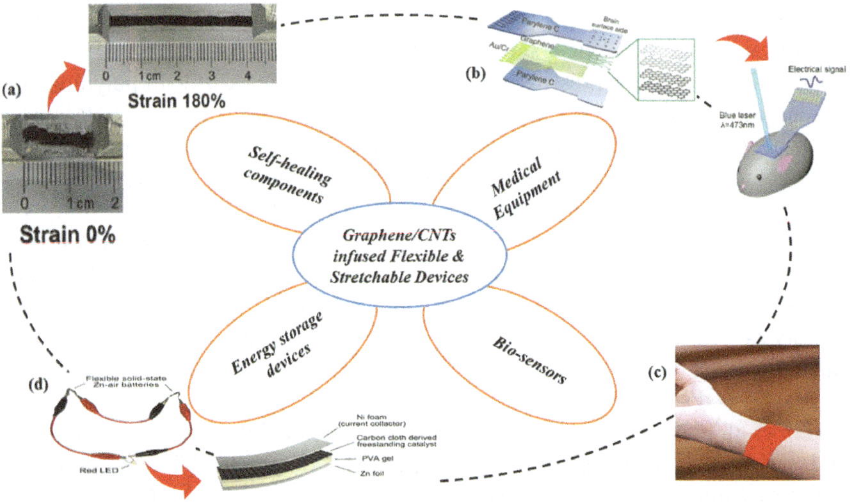

Fig. 1 Schematic illustration of the potential application areas for graphene and/or carbon nanotubes infused flexible and stretchable devices (**a**) flexible, stretchable and self-healing devices with CNT/graphene/PANI composite film [30], **b** Graphene-based transparent devices for optical imaging and optogenetics with carbo-layered electrode array and schematic representation of optogenetic testing [33], **c** a bandage strain sensor for observing human activities (like breathing and other motions) [21] and **d** photograph of two Flexible solid-state Zn–air batteries infused with carbon cloth derived free-standing catalyst and its schematic representation [29]

with deformable graphene photo-transducer) and their performance in colorimetric sensors.

The above review emphasized the importance of nano-scale materials (like graphene) for stretchable and flexible devices. Interestingly, the use of graphene nanoparticles (discovered in 2008) has reformed the entire technology, almost in every sector from highly-stretchable gauges to self-healable energy storage devices. The application summary of stretchable and flexible components with their area of application has been summarized in Fig. 1. Additionally, it is crucial to focus on the fundamental properties and uniqueness of graphene nanoparticles along with the new class of stretchable and flexible devices which has great potential to transform the existing technology.

2.1 Graphene: Background and Properties

Wallace in 1947 [34] explained the physical properties of 'graphene' through band-theory. However, by that time, graphene and its properties were part of the theoretical system. Nevertheless, the free-standing two-dimensional crystals of graphene layers are reported by Novoselov et al. [35] in 2004. For this discovery authors (Andre Geim, Konstantin Novoselov) received Nobel Prize (physics) in 2010. Graphene is extracted from graphite by mechanical exfoliation (repeated peeling) process by using the SiO_2 substrate. The extracted 2D film (or graphene) was found stable in the ambient atmosphere, and the size of a few layers of graphene is measured up to 10 μm. A unique property or physical behavior of graphene structure is the arrangement of carbon atoms in honeycomb lattice making 120^0 angle with adjacent carbon atoms with bonding length of 1.42 Å [36]. The carbon allotrope networks are stacked in AB fashion (Bernal stacking) and each sheet with 2D structure (Fig. 2a–b) is known as graphene, and the inter-sheet interaction is very week in nature [37]. The structure of graphene and its descendants (fullerenes, nanotubes and graphite) are shown in Fig. 2a with graphene structure with honeycomb arrangement and the lattice structure in Fig. 2b–c. Due to its unique design and nano-scale size; the graphene properties (mechanical, electrical, thermal and physiochemical) holds the exceptional potential for nano-technology-based future devices with ultra-efficient performance. Different physical properties of graphene are summarized in Table 1.

In recent days, the advancement in nano-technology (and their manufacturing methods) has dramatically increased the demand for efficient, flexible and stretchable miniaturized devices [42, 43]. The properties of graphene sheets such as larger surface area, high electrical and thermal conductivity, exceptional mechanical strength and flexible behavior make it best suitable material for building blocks for stretchable and flexible devices. It has also been noticed (*section: Flexible and Stretchable devices:*

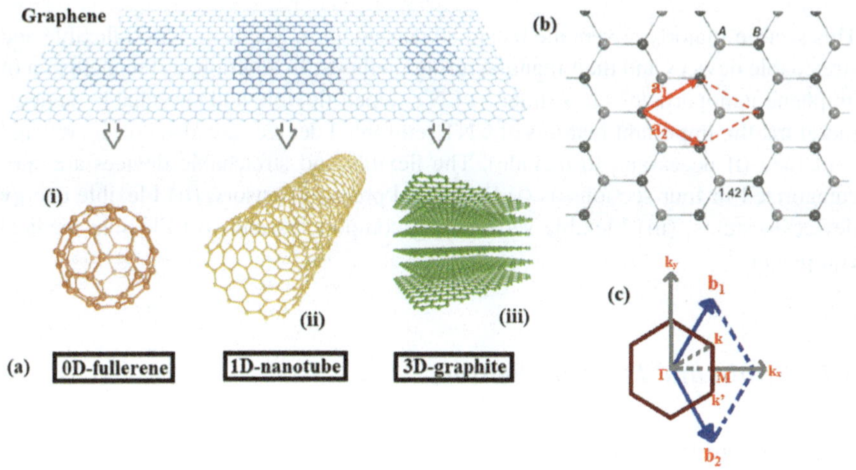

Fig. 2 Schematic illustration of **a** graphene and its descendants, (i) wrapped graphene or fullerenes with zero-dimension, (ii) rolled graphene or nanotubes with one-dimension and (iii) staked graphene or graphite with three-dimensional structure [38], **b** atomic structure (with honeycomb arrangement) of graphene in which grey colored sphere are carbon atoms *[the lattice can also be represented by two interpenetrated carbon triangular sublattices A (light grey) and B (dark grey)]*, additionally, the dotted blue lines denote the two-atom primitive cell with lattice vectors a1 and a2 and **c** The corresponding reciprocal lattice are represented by the red dotted lines with lattice vectors *b*1 and *b*2; the first Brillouin zone (BZ) is denoted by the hexagon with the high symmetry points Γ, *M*, and *K* (Dirac-point)

Table 1 Properties of graphene sheets [34, 35, 39–41]

Features	Parameters
Material	Semi-metal
Mechanical strength	~ 1 TPa
Young's modulus	1000 ± 100 (1050) GPa
Tensile strength	130 ± 10 (121) GPa
Charge currier mobility	~ 200,000 cm^2/Vs
Thermal conductivity	~ 5000 W/mK
Surface area	2630 m^2/gm

A brief history) that great efforts have been made to utilize the nanomaterials (like graphene) for designing a new class of graphene-based flexible devices for almost every sector. Interestingly, the development of flexible and stretchable devices needs a unique manufacturing strategy (recent advancement) which is another important aspect of the study, for the better understanding of potential application and utilization of these devices.

3 Graphene Infused Stretchable and Flexible Devices

This section majorly covers the recent advancement in graphene-based flexible and stretchable devices and their manufacturing methods. In addition, as the inclusion of graphene nanoparticles are similar to CNT doped flexible and stretchable devices; therefore, the important features of CNT reinforced devices are also considered and explained (if necessary, to include). The flexible and stretchable devices are sub-categorized in four sections as (i) Strain and pressure sensors, (ii) Flexible energy devices/systems, (iii) Flexible self-healing components and (iv) Flexible medical equipment.

3.1 Strain and Pressure Sensors

Strain and pressure gauges (or sensors) are those devices that measure strain (or deformation) and pressure (of liquids or gases) respectively whenever required. These sensors are the most common types of devices and used almost everywhere during experiments and/or physical measurements. However, in the present time, the demand for better wearable sensors have gained human attention for activity monitoring with accurate stretchable and flexible strain sensors. Yan et al. [44] defined the limitation of conventional strain gauge as they are suitable for low strain rates and mainly based on bulky technology. In the same research, they proposed an innovative method for the fabrication of graphene (3D macro-porous nano papers) and nanocellulose

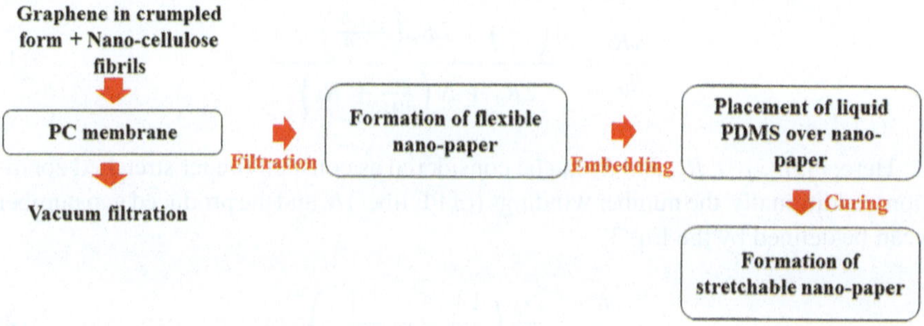

Fig. 3 **a** Schematic representation of the steps involved to fabricate the stretchable graphene nano-paper up to 100% stretching limit

infused highly stretchable strain sensors. Figure 3a represents the fabrication process of crumpled graphene and nanocellulose papers and then the addition of 3D paper sheets in the elastomer matrix to produce flexible strain sensors. The mixing ratio of graphene and nano-cellulose is 1: 1by weight and the fabrication process is based on solution-processable graphene technique.

The SEM scanning of the prepared sample showed the crumpled nano-paper as a porous structure. The comparative stretchable properties proved the importance of the fabrication method, as the flexible nano-paper offered flexibility up to 6% only. However, the graphene nano-paper infused in PDMS showed 100% elongation properties. The comparison (relative resistance properties and gauge factors as a function of strain %) of prepared stretchable strain with CNT and silver nanowires (AgNW) infused strain sensors [21, 45, 46] proved the significance of graphene embedded sensors over other nanoparticle infused sensors [44].

Graphene-coated compressible spring-like (fiber-based) composite sensors [47] exhibited exceptional performance under bending and torsion situations as a strain sensor. The fabrication process is covering the winding of double-covered yarn (polyurethane: PU core fiber and polyester: PE on the sleeve side) which is treated with air-plasma followed by fiber etching (PDCY). In the next step, dip-coating of graphene oxide (GO) is done, and in the final stage, PDCY-GO is soaked in hydroiodic (HI) acid for obtaining reduced GO (RGO) coated fiber which can be used in strain-sensor [47].

Cheng et al. [47] also focused on an essential aspect of resistance variation with varying strain rate from 0 to 50%. The SEM images highlighted that during the strain condition, the winding angle of PF and PU (core fiber) varies and a gap is created. Interestingly, the decrease in contact area (contact resistance) lead increase in contact resistance between inner and outer layers. Moreover, a mathematical model (Eq. 1) is proposed to calculate the relative resistance variation in terms of fiber radius (r_1 for PE fiber and r_2 for PU fibers respectively), the resistivity (ρ) and thickness (t) of RGO sheet, θ as winding angle, R_a is initial contact resistance between the inner and outer fibers.

$$\frac{\Delta R}{R_0} = \frac{n\left(\frac{\rho r_2}{t r_1}\right) + 2R_a\left(\frac{\sin 2\theta}{\sin 2\theta_0}\right) - 1}{2R_a + N\left(\frac{\rho r_1}{4 t r_2} + R_c\right)} \tag{1}$$

Here, r_1, r_2, ρ, t, R_c, and $R_a$can be considered as constant (under stretched condition); additionally, the number windings (of PE fiber) N and the produced gap number $n$can be defined by the Eq. 2.

$$\frac{n}{N} = \frac{2r_1}{L}\left(\frac{1+\in}{\cos\theta_0} - \frac{1}{\cos\theta}\right) \tag{2}$$

Therefore, it can be concluded that the variation in relative resistance depends on the generated gap number (n) and fiber winding (of PF fiber) number N.

Tang et al. [48] fabricated highly stretchable and ultra-sensitive strain gauges based on reduced graphene oxide microtubes. The stretching capacity of prepared sensors is reported 50% as compared to its original length with exceptional modulation capabilities depending upon the situation. The manufacturing method (template induced assembly followed PDMS polymer coating) is summarized in Fig. 4a–b. As a result, it can be concluded that the combination of easily sliding of RGO sheets over each-other coated with highly stretchable polymer PDMS based sensors are highly bendable and stretchable (Fig. 4c–d) up to 50% of the original length. In addition, these sensors offer high sensitivity of ~ 630-gauge factor, durable (Fig. 4e–f) and cost-effective as well. Pang et al. [49] proposed graphene network (porous-structure)

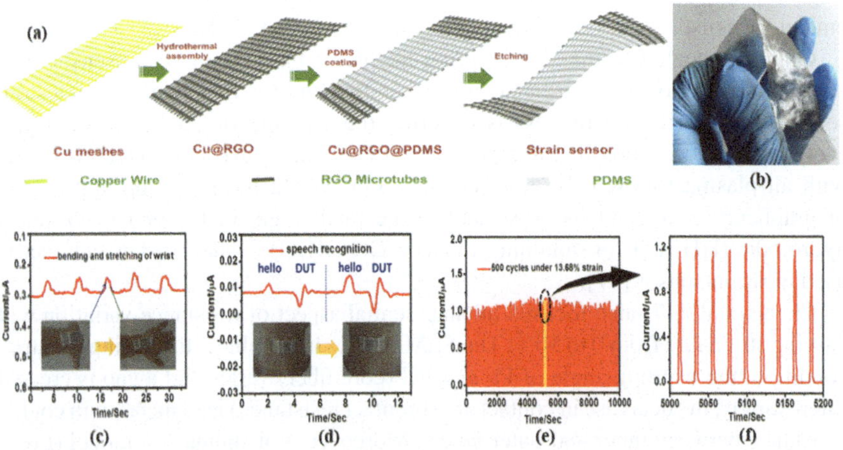

Fig. 4 **a** Fabrication steps (schematic representation) for RGO-microtubes elastomer strain sensors in which a copper mesh is used as a substrate material to deposit RGO followed by PDMS coating and etching process, **b** a digital photograph of prepared sensor and its flexibility, **c**–**d** performance (current vs time curves) of sensors under stretching of wrist and speaking different words, **e**–**f** durability tests at 13.68% strain condition up to 500 cycles and pattern of the curve in yellow-zone [48]

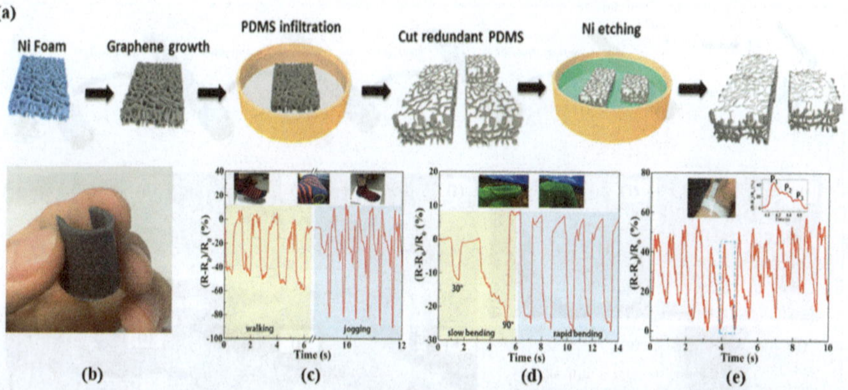

Fig. 5 **a** Fabrication process of pressure and strain sensors with the graphene porous network structures in which graphene growth took place over Ni-foam and pores infiltration with PDMS followed by bubble removal and removal of Ni-skeleton by hydrochloric acid, **b** a photograph, showing bending capabilities of GPN-PDMS composites, **c**–**e** Signal variations (relative-resistance) as a function of time during walking, bending and blood pressure variations in wrist respectively [49]

based flexible, stretchable, highly sensitive sensors for strain and pressure sensing devices for the first time. The fabrication process of flexible nickel-coated graphene sensors (with a photograph of GPN-PDMS composite) is shown in Fig. 5a–b. This research proved high potential for a variety of sensors (under multiple loadings such as walking, jogging, bending, etc.) by using graphene. The performance (resistance versus strain) of porous-graphene based sensor has two linear regions within strain range of 0–18 and 22–40% with the sensitivity 2.6 and 8.5 respectively. Therefore, these are better than normal pressure sensors. Figure 5c–e represents the sensor performance in practical human walking, finger-bending and wrist-pulse monitoring situations, respectively. Under multiple conditions, these foam-based sensors proved the highest-pressure range of 2000 kPa with 8.5 sensitivity which is the uniqueness of these sensors.

Gao et al. [50] in 2019 highlighted the challenges for developing flexible and breathable sensors (for full-body motions) under hazardous situations like corrosive atmosphere. Sensors with SiO_2/graphene shell and hierarchical TPU (thermoplastic polyurethane) core-based fibers (super-hydrophobic and conductive nanofiber composites) offer enhanced tensile strength, Young's modulus and elongation properties. The TPU nanofibers induced conductive and flexible composite membrane are manufactured by electrospinning method [51], which can offer ultra-hydrophobic and conductive features even after several stretch-cycles. Therefore, sensors have better mechanical and durability performance under wet conditions and/or non-favourable (salt, acid, or alkaline solution) situations. Figure 6a–i demonstrates the fabrication process of SiO_2/graphene/TPU sensors and their performance results including the performance in stretching and releasing conditions using the SEM analysis. In addition, Fig. 7a–h shows the testing response of prepared sensors under various

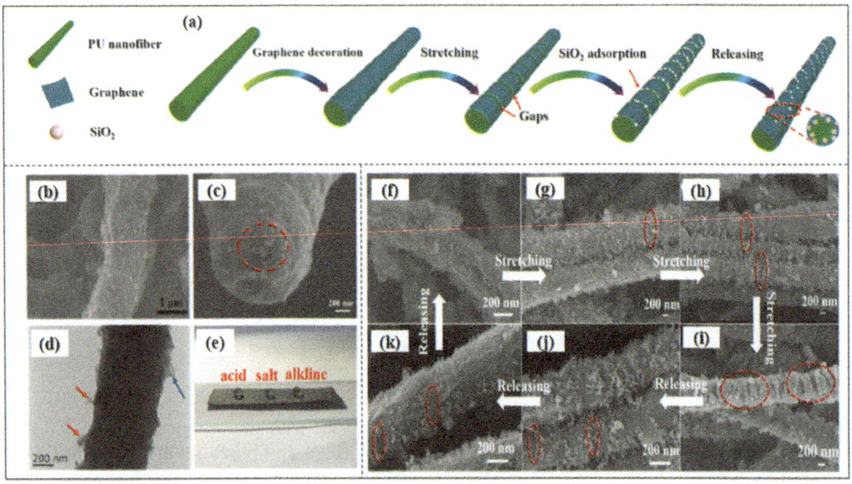

Fig. 6 **a** The schematic representation of the fabrication process of SiO_2/graphene/TPU sensors by using PU nanofibers manufactured by electrospinning method. **b–c** The surface morphology and cross-section of prepared fibers for sensors, **d** TEM image of fiber showing graphene shell (blue arrows) and nanoparticles (red arrows), **f–k** SEM images of super-hydrophobic and conductive nanofiber composites under stretching and releasing state ($\varepsilon = 10, 20, 50, 100, 50, 20\%$ respectively) [50]

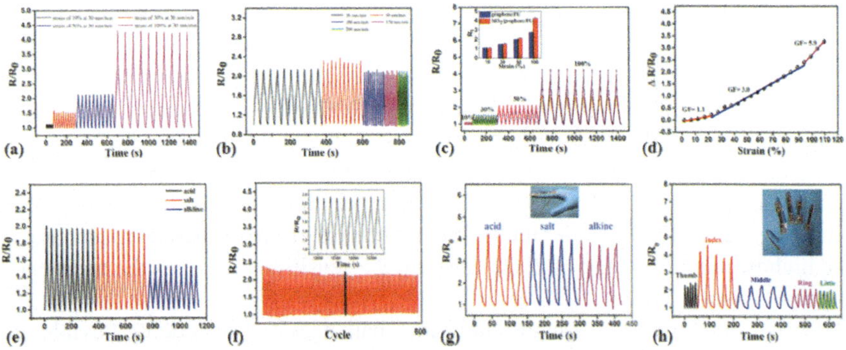

Fig. 7 **a** Resistance response (R/R0) as a function of time for strain sensor at the strain of 10, 30, 50 and 100% at 30 mm.min^{-1}, **b** the sensing performance of the strain sensor under various strain rate at 50% of strain, **c** the comparative analysis of resistance response graphs of graphene/PU-X strain sensor and SiO_2/graphene/PU-X strain sensor at 30 mm·min^{-1}, **d** representation of gauge factor (GF) of composite with different strain rate, **e** resistance response of the strain sensor after immersing it in acid, salt and alkaline solutions for six hours, **f** sensing performance of the fabricated strain sensor after 600 cycles of stretching and releasing and **g–h** performance of sensor after dipping finger in acid, salt and alkaline solution and movement of fingers and thumb [50]

conditions such as resistance and sensing performance under stretching conditions (or at different strain rates), gauge factor and influence of multiple solutions (acid, salt and alkaline) on the real-life application of fabricated sensors.

Flexible and stretchable sensors are not only limited toa specific body part. The Omni-purpose stretchable strain (OPSS) sensors proposed by Jeon et al. [52] have shown their significance for whole-body monitoring, including joint portions. The basic principle of these sensors involved the mechanism of nano-cracking structures which is inspired by the spider-sensing system. The fabrication of the OPSS sensors has been done by using a thermoplastic polyurethane-based solution (PU beads: (Pellethane 2363-80AE, Lubrizol), tetrahydrofuran: THF and dimethyl formamide: DMF) [53] by spin-coating method followed by magnetron sputtering method for the deposition of platinum (Pt) layer. A digital photograph and microscopic images (AFM and high-resolution SEM images) of OPSS sensor with crack-mechanism (at different density) are shown in Fig. 8a with its performance (Fig. 8b). These sensors are advantageous to monitor different functions related to whole-body movement and motions (Fig. 8c). Thus, the potential for crack-based sensors (like OPSS) has a wide-range of application for various functions.

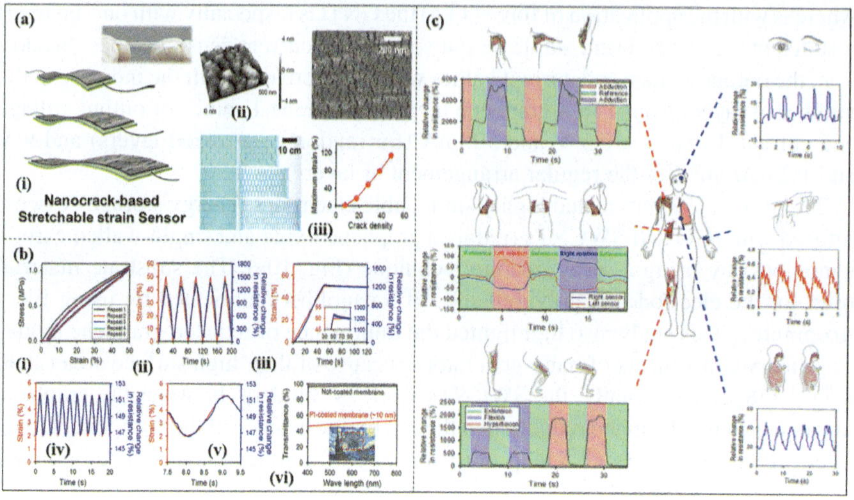

Fig. 8 **a** (i) A digital photograph and schematic representation of nano-cracking based omni-purpose stretchable strain (OPSS) sensor (ii) the atomic force microscopic image (3D) with high-resolution SEM of platinum layer (iii) paper model for showing the cracking structure (with different crack-density) at different failure strain, **b** mechanical and resistance performance of OPSS sensors (i) stress vs strain curves, (ii) change in resistance at cycling up to 50% of strain (iii) linearity performance with relative change in resistance at 50% strain condition (iv–v) relative change in resistance at 0.5 Hz frequency with 2 and 5% strain and close-up of single cycle pattern (vi) Transmittance spectra of the OPPS (platinum thickness ~ 10 nm) in visible wavelength (400–800 nm) Inset showing the sensor covering the middle portion of Vincent van Gogh's The Starry Night **c** Use of OPPS sensors for whole body functions monitoring (for example eye blinking, breathing, pulse-monitoring, knee, elbow motions like twisting and stretching etc.) [52]

3.2 *Flexible Energy Devices/Systems*

The energy field is one of the crucial sectors in present days; from mobile phones (in our hands) to satellites (in space) work on energy devices which make them functional. As per the report of the World Energy Council, the global energy demand will be double by 2060. Additionally, the environmental concerns have influenced researched to look for alternative options which are not derived by conventional sources like coal and petroleum products [54]. Interestingly, using non-conventional sources like solar and wind energy required lighter and flexible energy storage or generation devices. Therefore, the use of carbon-based nanomaterials (like graphene) has gained attention for supercapacitors, solar cells, batteries etc. as advanced energy devices [55].

Kim et al. [56] fabricated graphene embedded triboelectric nano-generators (GTNGs); the harvesting (large scale) of the graphene layer is done on copper/nickel layers by chemical vapour deposition (CVD) method. In this study, flexible and transparent GTNGs are prepared with one to four layers graphene harvested Cu, and Ni (with Bernal/rhombohedral stacking) foils. Fabrication steps are demonstrated in Fig. 9, in the absence of any external pressure or force the device is in neutral stage whereas with the application of force (1 kg) the GNTGs (especially with one-layered) resulted in highest and very stable output voltage and current density values. In addition, the output voltage and current values with are decreased with the increase in the number of layers from 1 to 4 graphene layers. The highest values of output voltage and current densities are 5 V and 500 nA (for randomly arranged layers) and 9 V and 1.2 $\mu A/cm^2$ (for the regular arrangement of layers).

Jeong et al. [57] fabricated highly stretchable electrodes (energy storage devices) infused with rGO and SWCNTs (reduced graphene oxide and single-walled carbon nanotubes) by using a spray coating technique (Fig. 10a). The substrate material used for the electrodes is polyurethane (PU) nanofibers fabricated by using Electrospinning. This study also highlighted the importance of reduced graphene oxides as primary constituents of nano-generators because of their high surface area (2630 m^2g^{-1} [58]). Additionally, the SWCNTs work as 'binders' between the graphene layers which results in the reduction in the internal resistance of the component [59]; interestingly, SWCNTs also help in re-stacking of graphene layers. The optimum

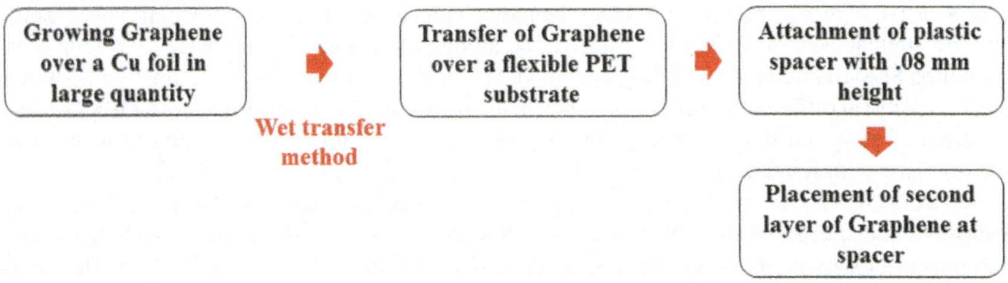

Fig. 9 Schematic illustration of the steps involved in the fabrication of one-layered graphene-coated and one-layered with PET/graphene layer at 0.8 mm spacer-distance based GTNG sensor

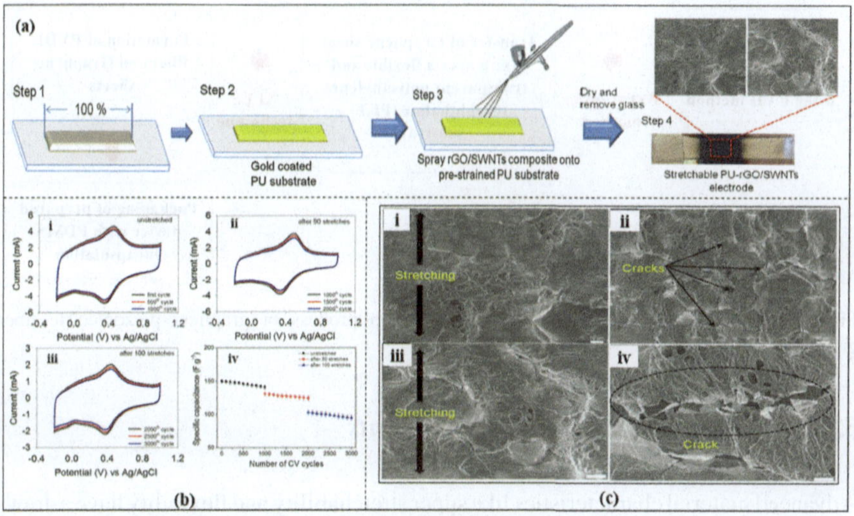

Fig. 10 **a** Fabrication steps of highly stretchable rGO/SWCNTs based electrodes by using the spray-coating method, **b** stability and specific capacitance performance of composite electrodes (i-iii) current as a function of potential in un-stretched, 50 and 100 stretching cycles respectively (iv) the comparative analysis of specific capacitance with respect of CV cycles number and **c** the SEM images showing the cracks generated which reduced the performance of rGO/SWCNTs based electrodes (i, iii) un-stretching, (ii, iv) stretching condition [57]

ratio of rGO and SWCNTs materials is 10:90, respectively, for best capacitive behavior after multiple stretching up to 100% strain. The highest measured value of capacitance is 265 F/g (for 1 M H_2SO_4 at 5 mV/s). Testing results also verified that ~ 65% of initial capacitance value retained (for unstretched) even after 100–3000 stretching cycles, as shown in Fig. 10b–c.

Fuh et al. [60] proposed a highly transparent and flexible piezoelectric generator embedded with the graphene layer. In this investigation, the graphene layer is deposited (by CVD method) on the Cu layer with the dimension of 2.5 × 2.5 cm. Later, the fabricated layer is transferred over a polyethylene terephthalate (PET) by wet-transfer method [61]. The piezoelectric polymers (PVDF) are placed (approximately 200 fibers in a parallel manner) by near-field Electrospinning (NFES) directly. In the final phase, the entire device is encapsulated in polydimethylsiloxane (PDMS), which is more durable and flexible. The fabrication process-steps of a flexible generator are shown in Fig. 11.

This study also showed that the graphene infused flexible piezoelectric generator offered better flexibility with transparent characteristics. The output voltage and currents values are measured as 2 V and 200 nA (for one-layered electrode) and 2.5 V and 250 nA (for two-layered electrode). Additionally, the performance of GPFGs under multiple flexible conditions (folding, diagonal folding and crumpling) are similar to results in normal conditions. The measured values for optical transmittance for single and two-layered electrodes are ~ 84 and ~ 81% respectively [60].

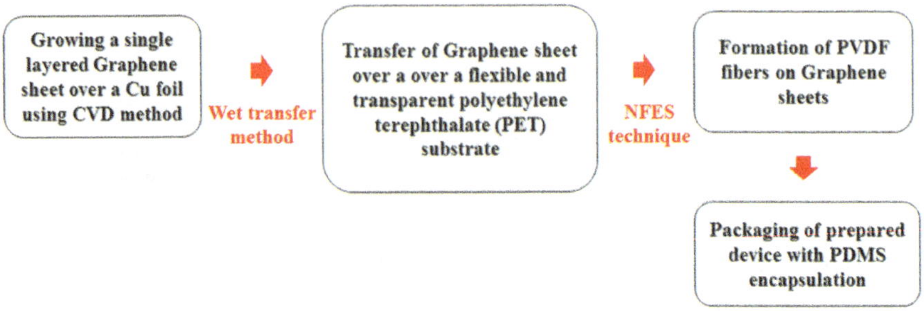

Fig. 11 Schematic representation of the fabrication process of graphene-piezoelectric fiber generator (GPFG)

3.3 *Flexible Self-Healing Components*

Advanced material characteristics like super stretchability and flexibility have a drawback of reduction in performance after a certain number of cycles. Surprisingly, naturally existing super flexible designs (like jellyfish)are capable of self-heal themselves [62]; this bio-inspiration can be a key to improve the durability of newly designed flexible and stretchable components. In 2012, Lu et al. [63] highlighted the problem of extended access lengths (ungated-channel region) in field-effect transistor: FET, which can be avoided by self-aligned top gates [64]. Therefore, the study performed by Lu and co-workers [63] proposed a fabrication method for self-aligned graphene layers-based (by CVD method) FET on a plastic substrate. In this transistor, the aluminium oxides are used as a gate and spacer material. Figure 12a (i) denotes the fabrication process of self-aligned graphene FETs additionally Fig. 12a (ii) shows

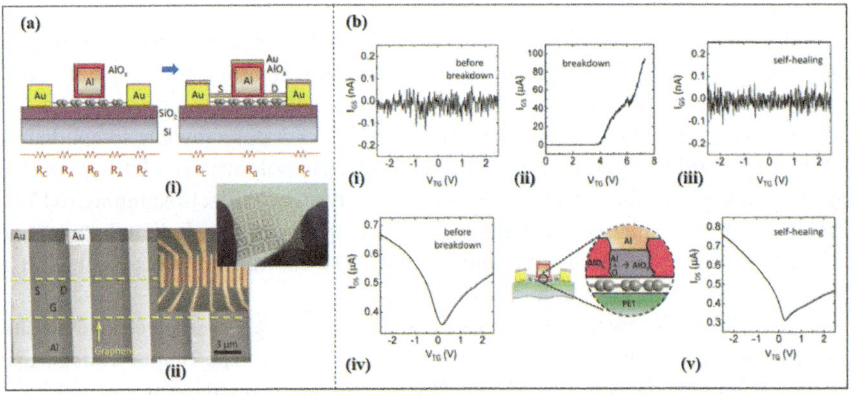

Fig. 12 **a** (i) Steps for the fabrication of self-aligned graphene-based field-effect transistors (ii) SEM image of prepared sensors with varying gate length, i.e. 300, 700 nm and 2 μm with a channel (of 10 μm width) inset: shows the overall device channels and flexible FET sensors, **b** the breakdown and self-healing performance of FET sensors [63]

the SEM image of fabricated FETs with varying gate length on Si substrate (inset is a device photograph). In the FET device, AlO_x is a vital material which offers high gate capacitance and provides the repairing (or self-healing) characteristics [65] during electrical breakdown. The fundamental mechanism of self-healing is known as electrical annealing or the ageing process in the presence of air.

Figure 12b (i–ii) shows the dielectric breakdown results during the current flow. Whereas, when the damaged oxides are exposed to the open-air environment for several hours (~ 6 h); it heals the damaged areas by newly generated oxides. Figure 12b (iii–v) shows the self-healing process and comparative performance of self-aligned graphene FETs which show the comparative results are significant. Cong et al. [66] proposed a novel class of self-healing graphene infused double network-based hydrogels with exceptional mechanical and self-repairing properties. The constituent materials of the hydrogel are graphene oxide (GO) and poly-acryloyl-6-aminocaproic acid (PAACA). In this study, the utilization of double cross-linkers (GO nano-sheets and calcium ions) technique has opted, which provided robustness in hydrogels. Additionally, the bonding network of prepared hydrogels is sandwiched by GO nano-sheets (containing oxygen molecule) and other groups of PAACA. Figure 13a–b represents the bond behavior of PAACA side chains and bond-mechanism of prepared GO-PAACA hydrogels with Ca^{2+} via coordination interaction [66].

The mechanical performance (tensile stress vs strain) of prepared PAACA and GO-PAACA hydrogels is shown in Fig. 13c for loading and loading–unloading conditions. However, the self-healing process is shown in Fig. 13d when just after adding a few drops of acid solution (at the raptured surface) and pressing at damage section can recover the hydrogel. Therefore, the possible area of this type of materials is drug delivery systems, biological scaffold and external coating [66]. In 2014, Kim et al. [67] highlighted the limited use of graphene infused flexible electronic devices due to lack of suitable deposition method and inherent defects. Therefore, they proposed a technique based on spray deposition technique by using a supersonic air-jet for reduced graphene oxide (r-GO) film. A schematic diagram of kinetic' spray injection-system' for r-GO suspension is shown in Fig. 14. The r-GO film coating over a glass substrate is done in three proportion, i.e. 10, 25 and 33% wt.% r-GO flakes; it is clear that the visibility (transmission reduced from 70 to 30% for 10–33% r-GO flakes) is reduced with the increasing concentration of r-GO flakes. The prepared r-GO coated substrates tested for flexibility aspects (manual bending), and it is observed that the LED (light-emitting diodes) continued to glow. Thus, it can be concluded that the electrical path of the coating doesn't break while bending, which signified the robustness of the prepared r-GO films.

Wang et al. [28] reported reduced graphene (RGO) induced highly flexible and self-healable springs (electrodes) with 295 μm diameter for supercapacitors. Figure 15a denotes the fabrication of spring-like structures (RGO based fibers) which are capable of stretching up to 300% and healable due to reconnection of tiny broken fibers in PU. Figure 15b indicates the preparation procedure of super-capacitor polypyrrole (PPy) embedded with RGO and MWCNTs fiber. In the first stage, a mixture containing GO, MWCNTs, SDS (sodium dodecyl sulphate) and VC

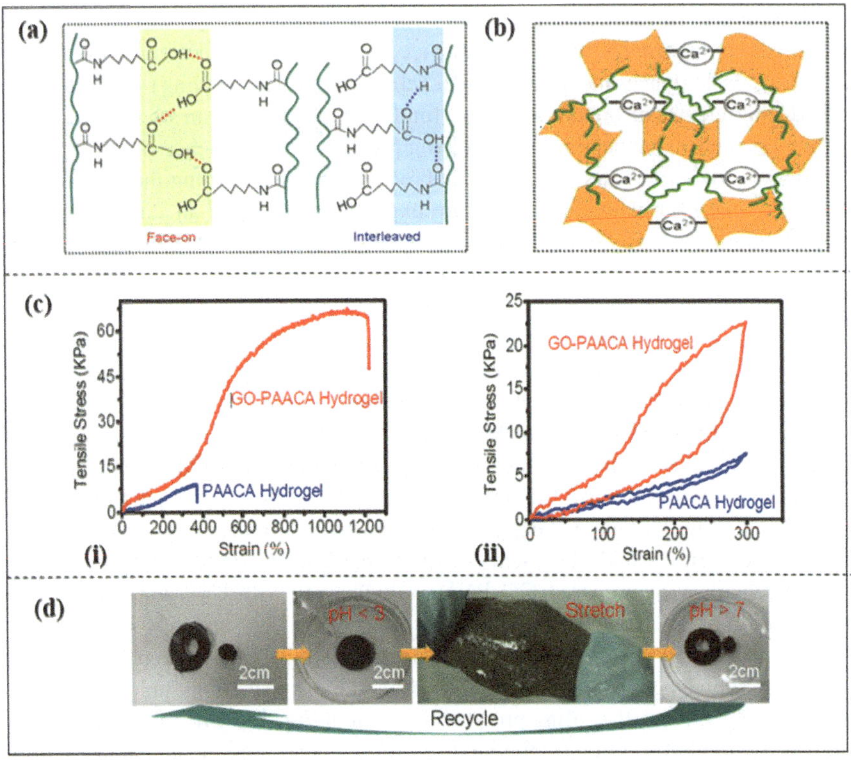

Fig. 13 **a** Representation of hydrogen-bonding with PAACA at low pH (green markings are backbone of a polymer chain with a face on Red and interleaved section: Blue), **b** the assembly mechanism of GO-PAACA hydrogels, **c** tensile performance of PAACA and GO-PAACA hydrogels (i) loading (ii) loading–unloading curves at 300% stretch condition, **c** Photographs representing the self-healing and recycling features of prepared hydrogels (GO-PAACA) [66]

(vitamin-C) is injected in a polytetrafluoroethylene pipe. Further, the tube is sealed at both ends using PDMS and cured in an oven at 90^0 C; this phenomenon reduced GO into RGO. After all necessary reactions [68] RGO/MWCNTs composite fibers are taken out and dried in normal room conditions. Afterwards, to improve the electrochemical and conductivity performance, a thin layer of PPy is deposited over semi-dried fibres and twisted in spring shape. Wrapping of twisted spring shaped fibers is done by poly(vinyl alcohol)-phosphoric acid (PVA-H_3PO_4) gel which can act as a separator and electrolyte as well. In the final stage, PU coating is done as an outermost layer which provides stretchable and self-healing characteristics to the capacitor with better mechanical properties.

Figure 15c i–ii indicated the stretchable (up to 300%) and self-healing properties of the fabricated supercapacitor. The photograph of the supercapacitor, electrochemical performance and applications of the supercapacitor is shown in Fig. 16a–i.

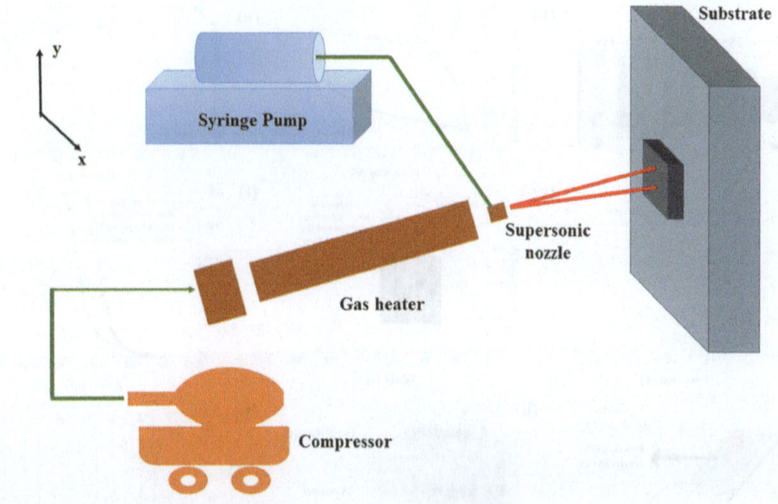

Fig. 14 Schematic diagram of the kinetic spray setup used for atomization and deposition of r-GO suspensions on substrate and TEM image of r-GO flakes (inset)

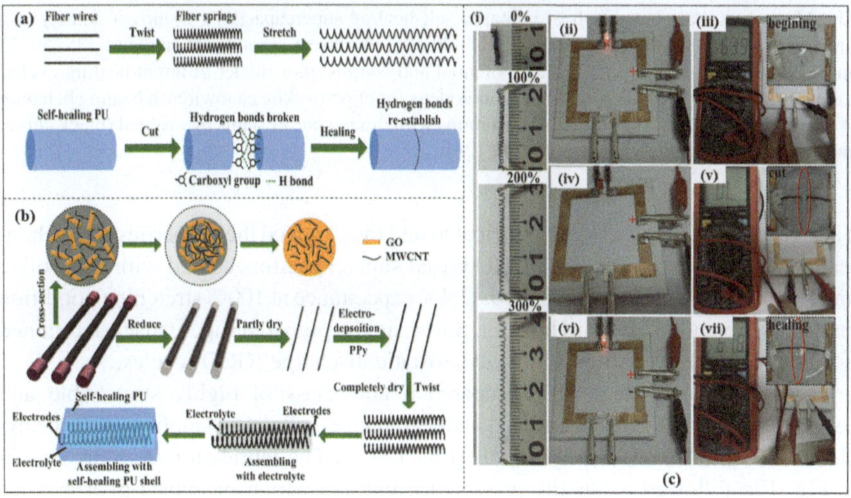

Fig. 15 **a** The illustration of fiber wires twisting in spring shape for making them stretchable and stretchable and self-healable PPy/RGO/MWCNTs electrodes after re-establishing hydrogen-bonds, **b** cross-sectional representation of GO/MWCNTs solution (injected in tubes) and when GO is reduced into RGO at 90 °C in oven then the RGO/MWCNTs wires are prepared and twisted. A layer of PPy is electrodeposited followed by PU shell coating for stretchable and self-healing properties, and **c** (i) representation of stretching characteristics of prepared spring shaped fibers up to 300% elongations (ii–iii) LED condition before the beginning of breaking, (iv-v) during circuit is broken and (vi-vii) after self-healing [28]

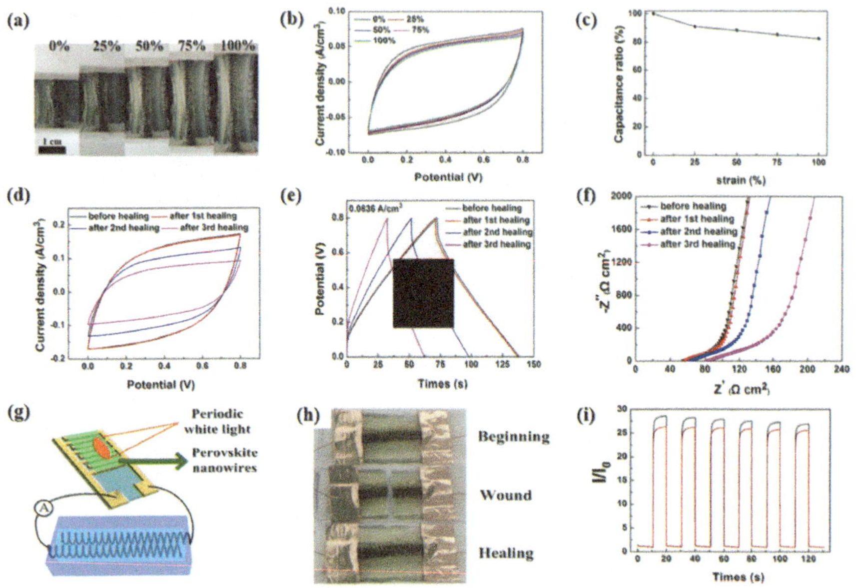

Fig. 16 **a** A digital photograph of prepared self-healing supercapacitors, **b** current density as a function of voltage at different stretching conditions, **c** value of capacitance ratio at 0, 25, 50, 75 and 100% strains, **d–f** current density, potential and Nyquist plots under different healing-cycles, **g** diagram of the supercapacitor driving a photodetector of perovskite nanowires, **h** healing behavior of supercapacitor and (i) photocurrent variations with time in on/off status by original (black curve) and self-healed (red curve) supercapacitor [28]

Therefore, authors successfully fabricated and investigated the properties of mechanically stretchable, self-healable RGO based supercapacitors coated with electrolyte and PU. The supercapacitor retained 82.4% capacitance at 100% stretching condition and 54.2% retention after 3rd healing. Surprisingly, prepared supercapacitor retained 84% capacitance even after 3000 galvanostatic discharge (GCD) cycles.

In 2019, Zheng et al. [69] prepared a new class of highly stretchable and self-healable strain sensors infused with graphene-cellulose nanofiber/poly (vinyl alcohol) (GN-CNF@PVA) hydrogels (Fig. 17a–f). The fabrication steps are shown in Fig. 17a followed a specific process for the fabrication of hybrid-hydrogels as initially, a gradual mixing of graphene-cellulose nanofiber is done to form a suspension. After mixing GN-CNF mixture is stirred and sonicated in room temperature conditions at 300 W and the aqueous suspension is evaporated up to 80 g of weight. The mixing ratio of GN-CNF is 0.15:1 (GN-CNF-A), 0.25:1 (GN-CNF-B) and 0.35:1 (GN-CNF-C). Simultaneously, poly-vinyl alcohol (PVA) (2 g) is dissolved in 20 mL deionized water at 90 °C and stirred until complete dissolution. After that, both solutions (GN-CNF and PVA) added slowly for 30 min at 90 °C. Finally, sodium tetraborate decahydrate (0.4 g) is dispersed into the mixture and when borax is completely dissolved; the solution is cooled to get hybrid-hydrogels (GN-CNF@PVA).

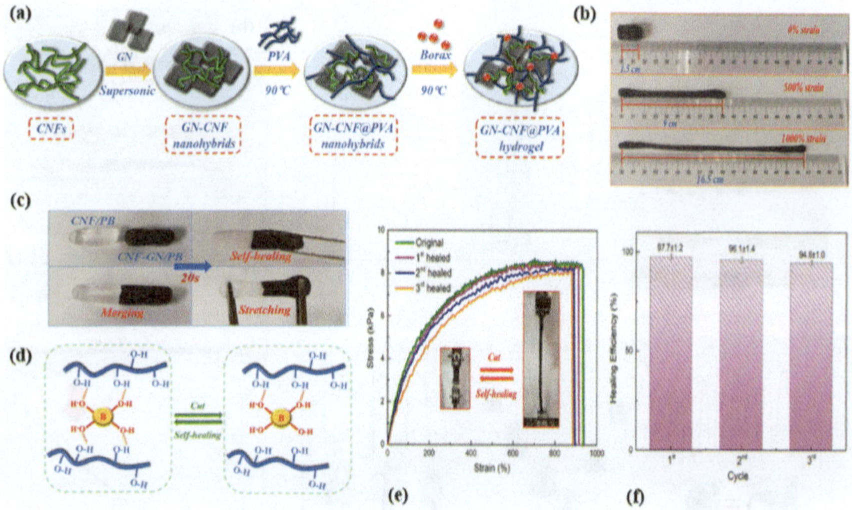

Fig. 17 **a** Preparation process of stretchable and self-healable strain sensors infused with graphene-cellulose nanofiber/poly (vinyl alcohol) (GN-CNF@PVA) hydrogels, **b** stretching abilities of prepared sensors up to 1000% strain, **c** self-healing of GN-CNF@PVA and CNF/PVA gels, **d** bonding mechanism of self-healing, **e** stress–strain behavior of hydrogels, i.e. original and up to 3rd healing cycle and **f** healing efficiency with respect to cycles [69]

The stretching capacity (up to 1000%) of GN-CNF@PVA-A gel is shown in Fig. 17b. CNF/PB and CNF/GN-PB offer self-healing behavior (as shown in Fig. 17b–c) after merging both. Figure 17e–f represents the mechanical behavior (stress–strain plots) and healing efficiency (after multiple cycles) respectively of GN-CNF@PVA hydro-gels. Zheng and co-authors [69] also performed the conductivity performance before and after self-healing, as shown in Fig. 18a. This figure indicates that the value of current changes during the cut/healing of GN-CNF@PVA-B hydrogels; however, current flow doesn't stop. Therefore, the use of these devices is best and suitable for self-repairing soft-electronics [70]. Similar to self-healing experiments, the prepared hydrogels are tested under different strain and flexible conditions (Fig. 18b–c), which proved the suitability of these strain-gauges for practical, flexible devices.

3.4 Flexible Medical Equipment

The human body is very complex and flexible; the complexity of body parts and structure can be understood by the motion of elbow, neck movement and movement of Adam's apple. In the past 20 years, developments related to monitoring human actions with flexible devices have gained sufficient attention [71, 72]. As a result, many soft and flexible devices (such as: endoscopic, and soft-robotic surgery tools)

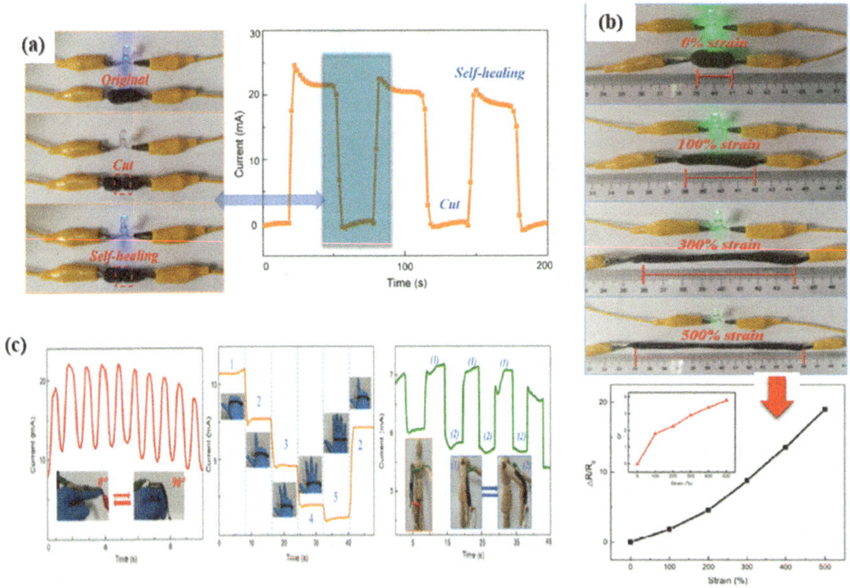

Fig. 18 **a** LED performance (current vs time curves) before, during and after self-healing, **b** LED illumination vs strain at 0, 100, 300 and 500% showing the reduction in illumination with an increase in strain, **c** current performance of prepared sensors under different hand-gestures [69]

are developed and used [73]; however, these devices are required many further modifications to make them painless and more human-friendly [74]. Remarkably, in the last decade, after the utilization of nanofibers, nanoparticles (CNTs, graphene etc.) to fabricate flexible and stretchable sensors have significantly changed the scenario related to human-friendly medical devices [75]. Yang et al. [76] proposed a new design of flexible triboelectric nano-generators (TENG) by using serpentine-pattern electrodes and Kapton-film (in wavy-structure). The uniqueness of this device is that it can be operated in stretching and compression modes even it can be comfortably attached with human-skin to monitor gentle motions of joints/muscles etc.

Yan et al. [77] fabricated highly stretchable graphene thermistors by a lithographic filtration method; the schematic representation and flexibilities aspects are shown in Fig. 19a (i–iii). Thermal detection channels are made by 3D crumpled graphene, whereas silver-nanowires (AgNWs) acted as electrodes. These thermistors are tested under different 0–50% of strains (shown in Fig. 19b) and 30–100°C temperature performance, as shown in Fig. 19c–h. Testing results proved a linear relationship curve ln*(R)* and 1000/T at strain rates between 0 and 50%. The change in electrical connection due to strain variation has an influence on the thermal index (B) which signifies the importance of thermistor for practical application related to human-body measurement. In addition, the speed-response behavior of thermistors is in good agreements in both relaxed and stretched conditions.

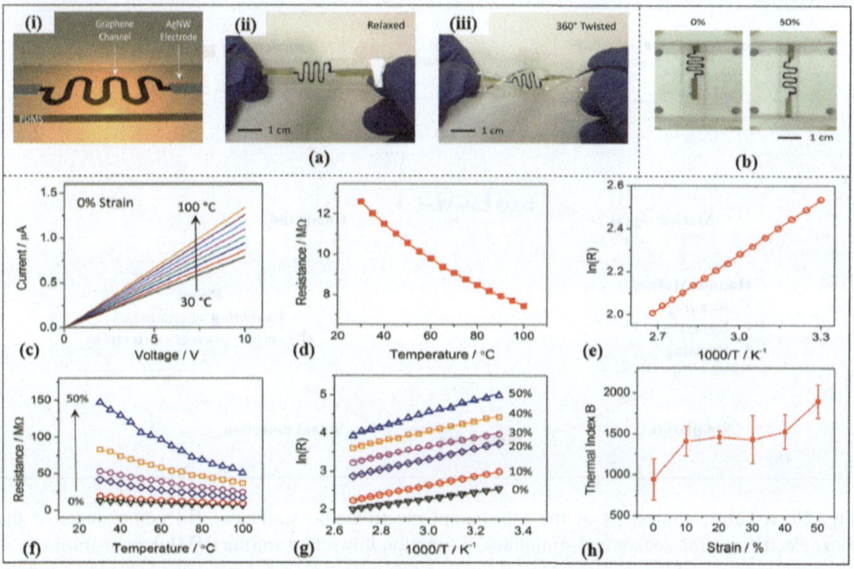

Fig. 19 **a** Representation of (i) different components (ii) relaxed position (iii) twisted position photograph of stretchable and highly conductive AgNWs based graphene thermistors, **b** strain (0 and 50%) conditions of the thermistor, **c** current vs voltage performance of thermistor ate 0% strain with varying temperatures from 30 to 100 °C, **d–e**) non-linear resistance variations with temperature and the linear relationship of ln(R) on 1000/T, **f** resistance variations with varying strain rates at different temperatures, **g** linear relationship of ln(R) on 1000/T at different strain values and **h** variation of the thermal index with varying strain [77]

Utilizing the benefits of graphene (GP), SWCNTs and silver nanowires (AgNWs), Lim et al. [78] developed stretchable and transparent motion sensors and interactable human–machine interface (iHMI) driven from patterned graphene heterostructures (transparent wearable electrotactile stimulator). These types of sensors and interactive robot-control technologies have a high potential for future medical applications. Figure 20a explains the manufacturing steps; (i), one graphene layer is transferred onto a substrate followed by a dispersion (for 30 s) of AgNWs onto the graphene layer and baked for 5 min at 200 °C. In the next step, the second layer of the graphene layer is transferred (on AgNWs embedded GP layer) which is isolated by reactive ion-itching method. The top GP-layer is patterned again by using the photolithography method, and a polymethylmethacrylate (PMMA) layer is spin-coated at the top for support purpose. The nickel layer is etched by nickel-etchant, and after this process, the simulator is transferred on PDMS/poly-vinyl alcohol: PVA film. In the final phase, PMMA film is removed by using acetone solution.

Kim et al. [79] implanted flexible and transparent bio-interface device infused with graphene-cell-sheets for monitoring and recording muscular and nerve movements. The detailed fabrication process of a cell-sheet-graphene hybrid device is illustrated in Fig. 21. The graphene layers are deposited by the CVD method on a copper-foil, and the Ni-layer is deposited over Si wafer. Subsequently, Cr and Au are both

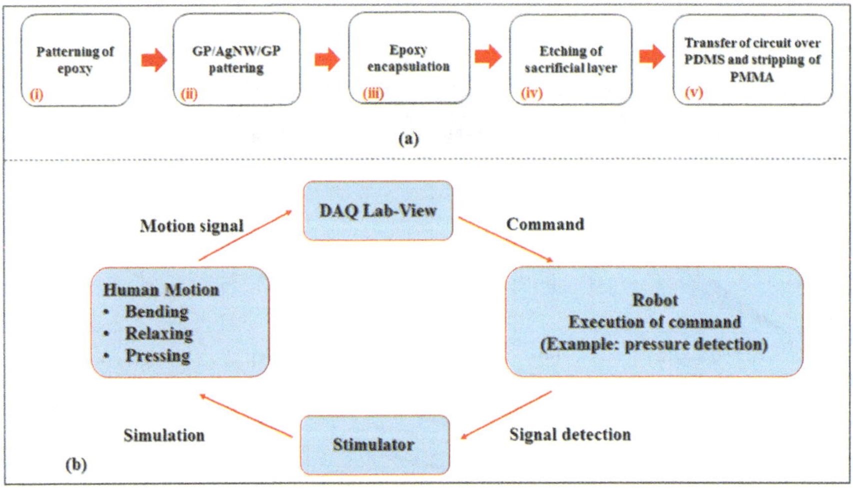

Fig. 20 **a** Fabrication steps of the transparent electrotactile stimulator (ii) components of the piezoelectric motion sensor and stimulator, **b** function flow-chart for the iHMI demonstration

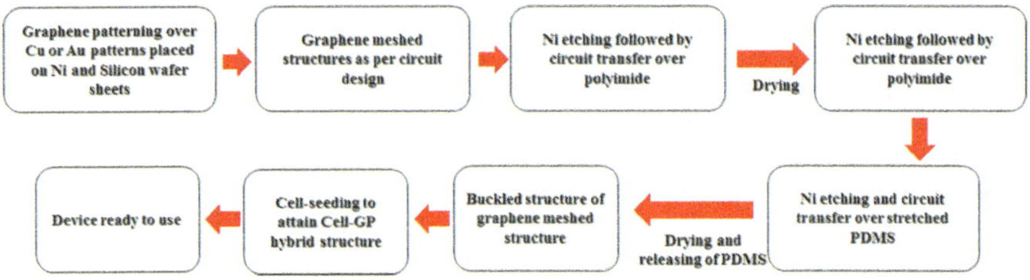

Fig. 21 The fabrication procedure and key applications in vitro and in vivo of the cell-sheet–graphene hybrid

deposited onto Ni-layer by using thermal evaporation technique. The patterning of Cr and Au layers is done by photolithography followed by the wet-etching method, and then the graphene sheet is transferred onto the sample and reinforced with $AuCl_4$ in nitromethane. In the final stage, the patterned sample is transferred to the polyamide layer which is coated on Ni/Si wafer, and then this is placed over 30% pre-stretched PDMS which is further dried for 10 min at 70°C. Pre-stretched PDMS sheets are slowly released; buckled graphene mesh is sterilized by ethanol (70% concentrated) for 15 min in the presence of UV light, and myoblast (C2C12) seeding (followed by culturing) is done [22, 80, 81].

This device contained a hybrid structure of cell-sheet-graphene, aligned C2C12 myoblast sheet, Au-reinforced graphene electrodes (mesh-patterned) mounted on a thick-polyamide membrane attached with soft PDMS substrate. The mechanical stretching and twisting showed the flexibility performance of the prepared hybrid device. This research proposed a new and useful strategy for medical implant multi-functional devices equipped with precise electrophysiological sensing, regeneration therapy using hybrid cells, optical and electrical simulations treatment. Additionally, these highly efficient, bio-compatible devices are soft and stretchable, which provides good mechanical stability. The performance of the implanted device in hind limb muscle of a mouse proved its performance for electrical and optical simulations in vivo and vitro (in monitoring C2C12 myoblast) opens new opportunities in soft bioelectronics. The negative response of the immune system for this type of implant is the most significant advantage from the perspective of its suitability with a living creature.

Park et al. [82] fabricated a graphene embedded skin-friendly sensor-array with microtopography guided conductive self-aligned patterns, as shown in Fig. 22a. The assembly of graphene nanoparticles is done on a liquid surface (in random-manner) by convective Marangoni self-assembly method (shown in Fig. 22a). Furthermore, this graphene nanoplatelet network (GNN) is transferred on a patterned stretchable PDMS substrate. The selected patterns are shown in Fig. 22b. In the final phase of the fabrication process, permanently sealing of two-sided conduction patterned surfaces is done by thin PDMS over-coating layers. The mechanical stretchability

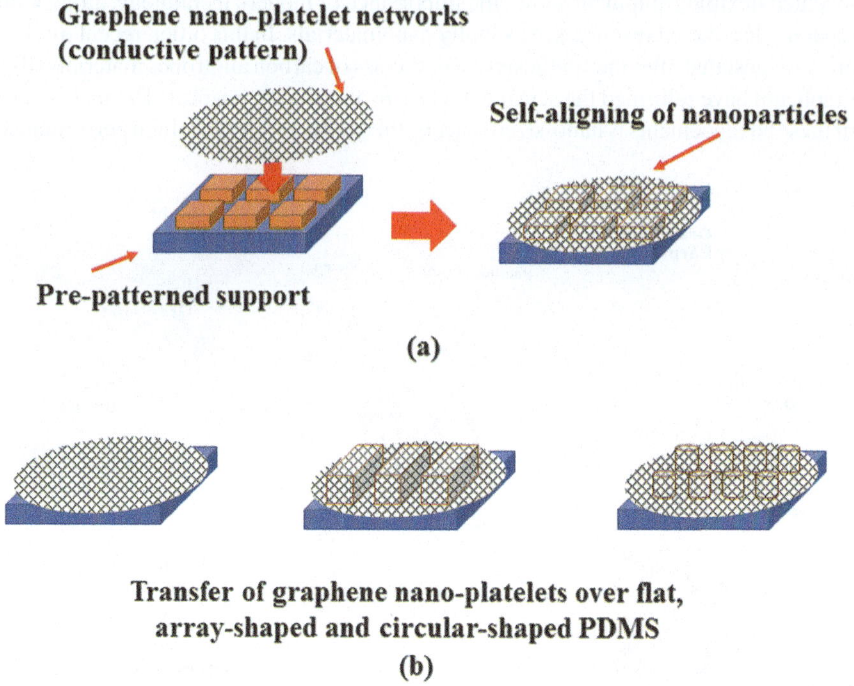

Fig. 22 **a** Schematic illustration of the fabrication process of GNN-infused conductive patterns by the liquid-driven transferring method and **b** conductive GNN patterns on PDMS substrate in various shapes (arrays and circular pillars)

of prepared sensors is <48% with the mechanical stability of 10^5 bending cycles, and the calculated sensitivity gauge factor is ~ 1697. The prepared wireless sensors are used to measure the amplified signals of different vital forces such as breathing, change in muscles tension, carotid pulse and radial artery. In addition, the waterproof coating makes these sensors potentially eligible to use it in artificial blood-vessels and during cardiovascular surgeries or other important biomedical applications as cost-effective devices.

4 Conclusions and Outlook

Humans have a lot of flexibility offered by different body parts (or senses) which perform numerous functions, and it is challenging to measure (or monitor) body functioning. In addition, all organs need different types of sensors; for example, a sensor which can be used at knee or elbow (required high bending features) needs extra flexibility as compared to a sensor which can be used on the neck (required high twisting characteristics). However, all these different body organs inspire to find and implement new equipment in some way (as shown in Fig. 23). Newly invented flexible equipment shows the importance of modern technology, along with recently discovered resources, especially nanomaterials. In this order, recent discoveries suggest that after the synthesis of a new class of carbon allotropes materials (like graphene) have reformed the sensing devices in almost every aspect. The main cause of these advancements is nano-sized (sheets, tubes structures) ultra-high performance

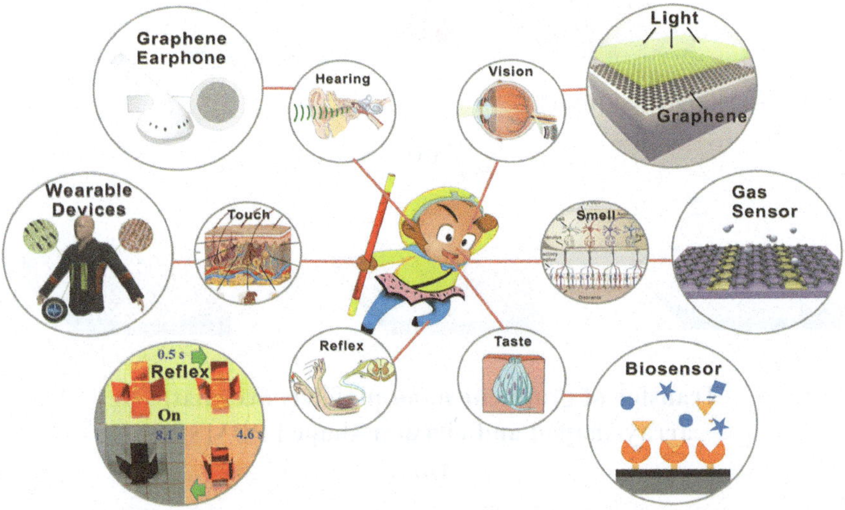

Fig. 23 Schematic representation of different flexible and stretchable graphene-based sensors (as per the human-body requirement) inspired by the human body [84]

(human-like sensing and reflex behaviour) components with better conductivity, pressure-sensing and/or self-healing nature. From the flexibility aspect, only nanoparticles are not sufficient, as they cannot produce devices ultimately. Other factors like substrate, sensors and covering constituents perform their role as a complete device. This chapter covered a detailed discussion of the fabrication process along with the utilization of flexible, stretchable and transparent materials as advanced sensors used in different areas. However, it is attractive to observe that, the fabrication process of almost every device followed a basic methodology, i.e., synthesis of a substrate and then the addition of other required elements over it. Therefore, the fabrication of substrate material plays an essential or fundamental role for these sensors. Most commonly adopted technique for fabricating substrate films is 'Electrospinning' which has covered a long journey with numerous modifications or advancements.

Nevertheless, it is the most adapted method for the fabrication of ultrathin films with nano-sized fibres arranged in specific directions. Ghosal et al. [83] have attractively described the history of electrospinning methodology and related the development phase with the present form of commercial technology. The brief definition of this method can be defined as, "Electrospinning is a method for fabricating ultrathin nano-scaled (diameter) fibres by melting a polymer solution which comes out from a 'spinneret' under a high voltage of electric field and collected (or solidifies) on collector (target) plate."

In this chapter, we have focused on the extraordinary journey and synthesis methodology of graphene nano-sheets. Due to their unique properties, they have been widely used as the main constituent of polymeric flexible and stretchable sensors. Interestingly, it has been noticed that with the use of graphene and polymer materials, any of sensor from strain-sensor to highly-sophisticated medical apparatus can be fabricated. These sensors are ultra light-weighted, high performance with unique flexibility, stretchability and transparent in appearance. The fabrication process of each sensor (strain, pressure, energy storage, medical and self-healing equipment) are different from each other depending upon the purpose of the application. As a conclusion point, it is essential to mention that all these developments have been made in the past decade; which indicates that fabricated gauges (discussed in this chapter) are not fully functional products.

Moreover, maximum sensors are prototypes of actual devices. It shows that these devices are still in the development phase, and during this development period optimization of design and functional properties is necessary. Therefore, it can be expected that the future of these flexile, stretchable and transparent human-friendly devices is very near; which will drastically change the existing scenario. We hope to get more human-friendly precise equipment, which can make human life more comfortable in every aspect of life and humankind. Additionally, we expect that it will become a reality soon.

References

1. Garnier, F. et al.: An all-organic" soft" thin film transistor with very high carrier mobility. **2**(12), 592–594 (1990)
2. Garnier, F., et al.: All-polymer field-effect transistor realized by printing techniques **265**(5179), 1684–1686 (1994)
3. Takei, K.: In: Flexible and Stretchable Medical Devices. Wiley (2018)
4. Yablonovitch, E., et al.: Extreme selectivity in the lift-off of epitaxial GaAs films **51**(26), 2222–2224 (1987)
5. Konagai, M., Sugimoto, M., Takahashi, K.J.J.o.c.g.: High efficiency GaAs thin film solar cells by peeled film technology. **45**, 277–280 (1978)
6. Hur, S.-H., et al.: Nanotransfer printing by use of noncovalent surface forces: applications to thin-film transistors that use single-walled carbon nanotube networks and semiconducting polymers **85**(23), 5730–5732 (2004)
7. Jiang, C., et al.: All-electrospun flexible triboelectric nano-generator based on metallic MXene nano-sheets **59**, 268–276 (2019)
8. Zhao, R., Lu, X., Wang, C.J.C.C.: Electrospinning based all-nano composite materials: recent achievements and perspectives **10**, 140–150 (2018)
9. Rogers, J.A. et al.: like electronic displays: Large-area rubber-stamped plastic sheets of electronics and microencapsulated electrophoretic inks. **98**(9), 4835–4840 (2001)
10. Someya, T. et al.: A large-area, flexible pressure sensor matrix with organic field-effect transistors for artificial skin applications. **101**(27), 9966–9970 (2004)
11. Menard, E., et al.: A printable form of silicon for high performance thin film transistors on plastic substrates **84**(26), 5398–5400 (2004)
12. Someya, T. et al.: Conformable, flexible, large-area networks of pressure and thermal sensors with organic transistor active matrixes. **102**(35), 12321–12325 (2005)
13. Ahn, J.-H., et al.: Heterogeneous three-dimensional electronics by use of printed semiconductor nanomaterials **314**(5806), 1754–1757 (2006)
14. Kim, K.S., et al.: Large-scale pattern growth of graphene films for stretchable transparent electrodes **457**(7230), 706 (2009)
15. Cao, Q., et al.: Medium-scale carbon nanotube thin-film integrated circuits on flexible plastic substrates **454**(7203), 495 (2008)
16. Sekitani, T., et al.: Stretchable active-matrix organic light-emitting diode display using printable elastic conductors **8**(6), 494 (2009)
17. Sekitani, T., et al.: A rubberlike stretchable active matrix using elastic conductors **321**(5895), 1468–1472 (2008)
18. Mannsfeld, S.C., et al.: Highly sensitive flexible pressure sensors with microstructured rubber dielectric layers **9**(10), 859 (2010)
19. Takei, K., et al.: Nanowire active-matrix circuitry for low-voltage macroscale artificial skin. **9**(10), 821 (2010)
20. Kim, D.-H., et al.: Dissolvable films of silk fibroin for ultrathin conformal bio-integrated electronics **9**(6), 511 (2010)
21. Yamada, T., et al.: A stretchable carbon nanotube strain sensor for human-motion detection **6**(5), 296 (2011)
22. Han, T.-H., et al.: Extremely efficient flexible organic light-emitting diodes with modified graphene anode **6**(2), 105 (2012)
23. Chae, S.H. et al.: Transferred wrinkled Al_2O_3 for highly stretchable and transparent graphene–carbon nanotube transistors. **12**(5), 403 (2013)
24. Zang, J., et al.: Stretchable and high-performance supercapacitors with crumpled graphene papers **4**, 6492 (2014)
25. Xu, Y., et al.: Flexible. Stretchable, and rechargeable fiber-shaped zinc-air battery based on cross-stacked carbon nanotube sheets **54**(51), 15390–15394 (2015)
26. Lee, H., et al.: Flexible and stretchable optoelectronic devices using silver nanowires and graphene **28**(22), 4541–4548 (2016)

27. Chiang, C.-W., et al.: Highly stretchable and sensitive photodetectors based on hybrid graphene and graphene quantum dots **8**(1), 466–471 (2016)
28. Wang, S., et al.: Highly stretchable and self-healable supercapacitor with reduced graphene oxide based fiber springs. **11**(2), 2066–2074 (2017)
29. Wang, H.-F. et al.: Defect-rich carbon fiber electrocatalysts with porous graphene skin for flexible solid-state zinc–air batteries. **15**, 124–130 (2018)
30. Liang, X., et al.: A dynamic stretchable and self-healable supercapacitor with a CNT/graphene/PANI composite film **10**(47), 22329–22334 (2018)
31. Shin, H., et al.: Stretchable electroluminescent display enabled by graphene-based hybrid electrode **11**(15), 14222–14228 (2019)
32. Snapp, P. et al.: Colloidal photonic crystal strain sensor integrated with deformable graphene phototransducer. 1902216 (2019)
33. Kim, T., Cho, M., Yu, K.J.J.M.: Flexible and stretchable bio-integrated electronics based on carbon nanotube and graphene **11**(7), 1163 (2018)
34. Wallace, P.R.J.P.R., The band theory of graphite. 1947. **71**(9): p. 622.
35. Novoselov, K.S., et al.: Electric field effect in atomically thin carbon films **306**(5696), 666–669 (2004)
36. Pinto, H.P., Leszczynski, J.J.H.o.c.n.m.: Fundamental properties of graphene. **5**, 1–38 (2014)
37. Kelly, B.J.L.N.J.: Physics of graphite. Appl. Sci. 267 (1981)
38. Crespi, V.H., et al.: Prediction of a pure-carbon planar covalent metal **53**(20), R13303 (1996)
39. Petroski, H.: The pencil: A history of design and circumstance. 1992: Alfred a Knopf Incorporated (1992)
40. Lee, C., et al.: Measurement of the elastic properties and intrinsic strength of monolayer graphene **321**(5887), 385–388 (2008)
41. Liu, F., Ming, P., Li, J.J.P.R.B.: Ab initio calculation of ideal strength and phonon instability of graphene under tension. **76**(6), 064120 (2007)
42. Wu, C., et al.: Two-dimensional vanadyl phosphate ultrathin nano-sheets for high energy density and flexible pseudocapacitors **4**, 2431 (2013)
43. Lu, Q., et al.: Free-standing carbon fiber cloth/sulfur composites for flexible room-temperature sodium-sulfur batteries **8**, 77–84 (2017)
44. Yan, C. et al.: Highly stretchable piezoresistive graphene–nanocellulose nanopaper for strain sensors. **26**(13), 2022–2027 (2014)
45. Xu, F., Zhu, Y.J.A.m.: Highly conductive and stretchable silver nanowire conductors. **24**(37), 5117–5122 (2012)
46. Lipomi, D.J., et al.: Skin-like pressure and strain sensors based on transparent elastic films of carbon nanotubes **6**(12), 788 (2011)
47. Cheng, Y. et al.: A stretchable and highly sensitive graphene-based fiber for sensing tensile strain, bending, and torsion. **27**(45), 7365–7371 (2015)
48. Tang, Y. et al.: Highly stretchable and ultrasensitive strain sensor based on reduced graphene oxide microtubes–elastomer composite. **7**(49), 27432–27439 (2015)
49. Pang, Y. et al.: Flexible, highly sensitive, and wearable pressure and strain sensors with graphene porous network structure. **8**(40), 26458–26462 (2016)
50. Gao, J. et al.: Electrically conductive and fluorine free superhydrophobic strain sensors based on SiO_2/graphene-decorated electrospun nanofibers for human motion monitoring. **373**, 298–306 (2019)
51. Huang, X. et al.: Stretchable, electrically conductive and superhydrophobic/superoleophilic nanofibrous membrane with a hierarchical structure for efficient oil/water separation. **70**, 243–252 (2019)
52. Jeon, H., et al.: Omni-purpose stretchable strain sensor based on a highly dense nanocracking structure for whole-body motion monitoring **9**(48), 41712–41721 (2017)
53. Cho, S.J., et al.: A rubberlike stretchable fibrous membrane with anti-wettability and gas breathability **23**(45), 5577–5584 (2013)
54. Jun, H.K.: Introduction to nanomaterials in energy devices. In: Nano-materials in Energy Devices, pp. 1–6. CRC Press (2017)

55. Grande, L. et al.: Graphene for energy solutions and its printable applications. 191–236 (2015)
56. Kim, S., et al.: Transparent flexible graphene triboelectric nano-generators **26**(23), 3918–3925 (2014)
57. Jeong, H.T. et al.: Highly stretchable reduced graphene oxide (rGO)/single-walled carbon nanotubes (SWNTs) electrodes for energy storage devices. **163**, 149–160 (2015)
58. Cheng, Q., et al.: Graphene and carbon nanotube composite electrodes for supercapacitors with ultra-high energy density **13**(39), 17615–17624 (2011)
59. Fan, Z., et al.: A three-dimensional carbon nanotube/graphene sandwich and its application as electrode in supercapacitors **22**(33), 3723–3728 (2010)
60. Fuh, Y.K., et al.: A transparent and flexible graphene-piezoelectric fiber generator **12**(14), 1875–1881 (2016)
61. Suk, J.W., et al.: Transfer of CVD-grown monolayer graphene onto arbitrary substrates. **5**(9), 6916–6924 (2011)
62. Cao, Y., et al.: Self-healing electronic skins for aquatic environments **2**(2), 75 (2019)
63. Lu, C.-C., et al.: High mobility flexible graphene field-effect transistors with self-healing gate dielectrics **6**(5), 4469–4474 (2012)
64. Farmer, D.B., Lin, Y.-M., Avouris, P.J.A.P.L.: Graphene field-effect transistors with self-aligned gates. **97**(1), 013103 (2010)
65. Miyazaki, H. et al.: Resistance modulation of multilayer graphene controlled by the gate electric field. **25**(3), 034008 (2010)
66. Cong, H.-P., Wang, P., Yu, S.-H.J.C.o.M.: Stretchable and self-healing graphene oxide–polymer composite hydrogels: a dual-network design. **25**(16), 3357–3362 (2013)
67. Kim, D.Y., et al.: Self-healing reduced graphene oxide films by supersonic kinetic spraying **24**(31), 4986–4995 (2014)
68. Qu, G., et al.: A fiber supercapacitor with high energy density based on hollow graphene/conducting polymer fiber electrode **28**(19), 3646–3652 (2016)
69. Zheng, C., et al.: Highly stretchable and self-healing strain sensors based on nanocellulose-supported graphene dispersed in electro-conductive hydrogels **9**(7), 937 (2019)
70. Tong, X. et al.: Swelling and mechanical behaviors of carbon nanotube/poly (vinyl alcohol) hybrid hydrogels. **61**(8–9), 1704–1706 (2007)
71. Baek, J.-Y., et al.: Flexible polymeric dry electrodes for the long-term monitoring of ECG **143**(2), 423–429 (2008)
72. Xu, S. et al.: Soft microfluidic assemblies of sensors, circuits, and radios for the skin. **344**(6179), 70–74 (2014)
73. Degani, A. et al.: Percutaneous intrapericardial interventions using a highly articulated robotic probe. In: The First IEEE/RAS-EMBS International Conference on Biomedical Robotics and Biomechatronics, BioRob 2006. IEEE, (2006)
74. Kim, Y.-J., et al.: A stiffness-adjustable hyperredundant manipulator using a variable neutral-line mechanism for minimally invasive surgery **30**(2), 382–395 (2013)
75. Ko, G. et al.: Flexible/stretchable devices for medical applications (2018)
76. Yang, P.K. et al.: A flexible, stretchable and shape-adaptive approach for versatile energy conversion and self-powered biomedical monitoring. **27**(25), 3817–3824 (2015)
77. Yan, C., Wang, J., Lee, P.S.J.A.n.: Stretchable graphene thermistor with tunable thermal index. **9**(2), 2130–2137 (2015)
78. Lim, S., et al.: Transparent and stretchable interactive human machine interface based on patterned graphene heterostructures **25**(3), 375–383 (2015)
79. Kim, S.J. et al.: Stretchable and transparent biointerface using cell-sheet–graphene hybrid for electrophysiology and therapy of skeletal muscle. **26**(19), 3207–3217 (2016)
80. Kim, S.J. et al.: Multifunctional cell-culture platform for aligned cell sheet monitoring, transfer printing, and therapy. **9**(3), 2677–2688 (2015)
81. Bae, S. et al.: Roll-to-roll production of 30-inch graphene films for transparent electrodes. **5**(8), 574 (2010)
82. Park, Y., et al.: Microtopography-guided conductive patterns of liquid-driven graphene nanoplatelet networks for stretchable and skin-conformal sensor array **29**(21), 1606453 (2017)

83. Ghosal, K. et al.: Electrical spinning to electrospinning: a brief history. (2018)
84. Zhang, Q., et al.: Human-like sensing and reflexes of graphene-based films **3**(12), 1600130 (2016)

Dimension Controlled Polymeric Matrices and Graphene Filler-Based Nanofibres by Electrospinning

Sasmita Mishra, Ajeet Singh, and Sandip Singh

Abstract Polymer graphene-based nanofibres obtained using electrospinning are increasingly used in wound dressing materials, filtration membranes, smart mats, and catalytic supports for catalysis, energy harvesting, conversion, and storage, photonic/electronic devices and various biomedical applications and thus attracting more and more research in the material chemistry. After keeping the importance of polymer graphene-based nanofibres in mind, first part of this chapter deals with the methods used so far to control the dimension of the polymer matrix by using different published articles. Second part of this chapter briefs about the properties of the electrospun nanofibres which showed better performances after the incorporation of graphene or graphene oxide (GO) filler and their applications.

Keywords Polymer graphene-based nanofibres · Dimension controlled polymer matrix · Electrospinning · Graphene filler-based nanofibres

1 Introduction

In 1600, William Gilbert observed the cone-shaped water droplets formed in the electric field presence brought the electrospinning concept into the picture [1]. In the year 1747, Abbeì Nollet showed that when water passes through an electrostatically charged vessel, it formed an aerosol [2]. Later, Lord Rayleigh studied the behaviour of charged drops [3]. As we know, a high voltage is required to eject liquid jets in both electrospinning and electrospraying technique. The only difference is that in electrospinning, continuous jet forms fibres. In contrast, in electrospraying, the jet breaks to droplets to form particles. The difference lies due to the viscosity of the liquid. In another sense, electrospinning is another form of electrospraying [4]. In general,

S. Mishra · A. Singh (✉)
Discipline of Chemistry, Indian Institute of Technology Indore, Simrol, Khandwa Road, Indore 453552, India

S. Singh
Catalysis and Inorganic Chemistry Division, CSIR-National Chemical Laboratory, Dr. Homi Bhabha Road, Pune 411008, India

S. K. Tiwari et al. (eds.), *Electrospinning of Graphene*, Carbon Nanostructures,
https://doi.org/10.1007/978-3-030-75456-3_3

different electrospinning techniques known are co-axial electrospinning [5], emulsion electrospinning [6] and melt electrospinning [7]. The co-axial electrospinning technique spuns two or more polymer solutions using co-axial capillaries to form the continuous coated or hollow nanofibres [5, 8]. In an emulsion electrospinning technique, electrospinning of emulsions either water-in-oil (W/O) or oil-in-water (O/W) is carried out to encapsulate hydrophilic or hydrophobic compounds which help in the fabrication of core–shell nanofibres [6]. And the melt electrospinning is a solvent-free method in which the polymer melt jet is used to produce nanofibre [9].

So far, various techniques are known for the nanofibres production, which includes template synthesis, drawing, self-assembly, phase separation and most important electrospinning [10–14]. Electrospinning is the best technique as it can produce a range of nanofibres according to their size of micro- or nanoscale [14] using the electrostatic field on a jet of a polymer solution [15, 16].

In recent years, electrospinning has been an attractive technique to fabricate polymer-based nanofibres. An electrospin setup has three components which are a high voltage source or supplier, a tube (needle) or capillary of small diameter, and a metal collector. The jet from the needle is governed by the electrostatic field and gets elongated in the fibres form after the evaporation of the polymer–solvent [17]. We have discussed the electrospinning and their types in our previous chapter of this book. Here, the effect of the design of different spinneret and collectors is present in brief (Fig. 1).

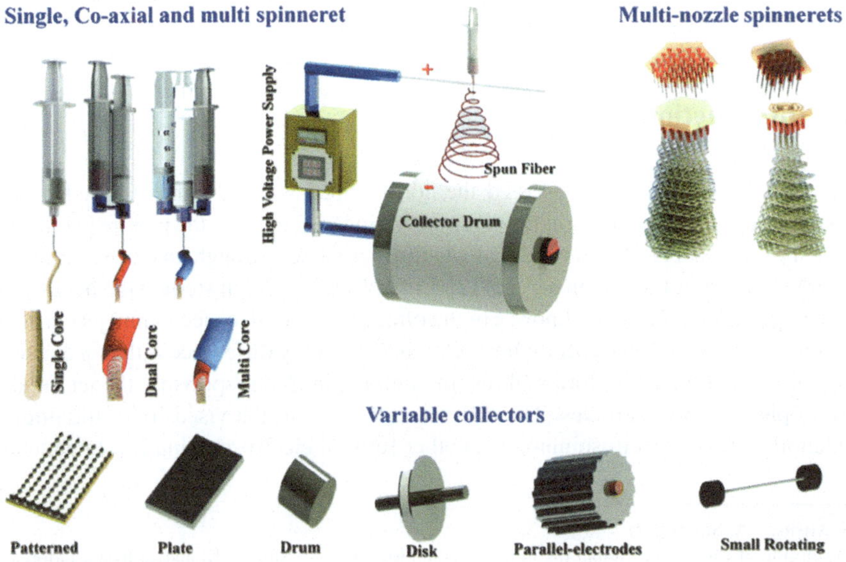

Fig. 1 Schematic representation of the electrospinning setup with different spinnerets, electrospinning setup and various collectors. Reproduced from Ref. [18] with permission from The Royal Society of Chemistry

Figure 1 shows three types of electrospinning based on the type of spinnerets. The single-spinneret-based electrospinning uses a needle which contains a polymer solution. It can produce single-core electrospun fibres. Co-axial electrospinning uses spinneret which has two co-axial capillaries which contain a polymer solution (for shell), and compound solution (for core) can be used to produce a dual-core fibre. Herein, the outer and inner precursor solutions after reaching to the nozzle mix and form a composite droplet. Due to the presence of charges on their surface, generation of viscous stress takes place first in the shell solution followed by the core solution, to generate a co-axial fibre. Similar to co-axial electrospinning, if multiple spinnerets will be used it will form a multi-core fibre.

The right side of the figure shows multiple nozzle spinnerets; such systems can be used to produce large quantities of fibres at a time. In such type of system, a strong repulsion can reduce the fibre production rate; result in a poor quality of fibre, also non-uniform fibre deposition [19]. So, the needles need to be arranged at an appropriate distance, and for the vast production of fibres, a considerable space will be required to accommodate such setup. Electrospinning set up can also be varied by using different types of collectors. A collector is a base having an opposite charge to accelerate the fibre towards itself. Aluminium foil is the ideal electrode material used as collector due to its lower cost, high availability, and easy replacement. This collector is connected to a ground electrode which allows the current to pass through. They are different types; some of them are mentioned as follows.

Patterned electrodes are a derivation of the parallel electrode geometry. In this system, more electrodes are used to obtain multi-layered mats with controlled hierarchical structures.

The plate or frame electrodes can also be used as a collector; different materials can produce different alignments of the electrospun fibre, for example an aluminium rectangular shape plate or frame develops more aligned fibres than a wooden one. Various parameters can affect the alignment of the fibres, which includes frame inclination angle, the shape and size, and the space between jet and the collector [20].

A rotating drum electrode is a collector which helps in the production of aligned fibres [20] than random fibres. This type of collector rotates at speed called an alignment speed (jet deposition and linear velocity of the drum surface matches).

A rotating disc collector directs the electrostatic field lines on the sharp edge of the disc, along with the rotation of the disc. This collector produces properly aligned fibres. However, the electrospun fibres have a small area of fibre deposition, and the alignment may be lost as the mesh extends in the 3D direction.

Parallel electrodes generally use two parallel grounded plates, rods or wires. Due to such arrangement, the jet stretching takes place which produces aligned electrospun fibres. In this setup, the stretching of the jet across the gap between the parallel conducting collectors attracts the field lines and hence, influences the electrospinning method [12].

Electrospun nanofibres are getting much attention also due to their applications as wound dressing materials [21], as filtration membranes [22], even smart mats [23], catalytic supports for catalysis [24, 25], energy harvesting, conversion,

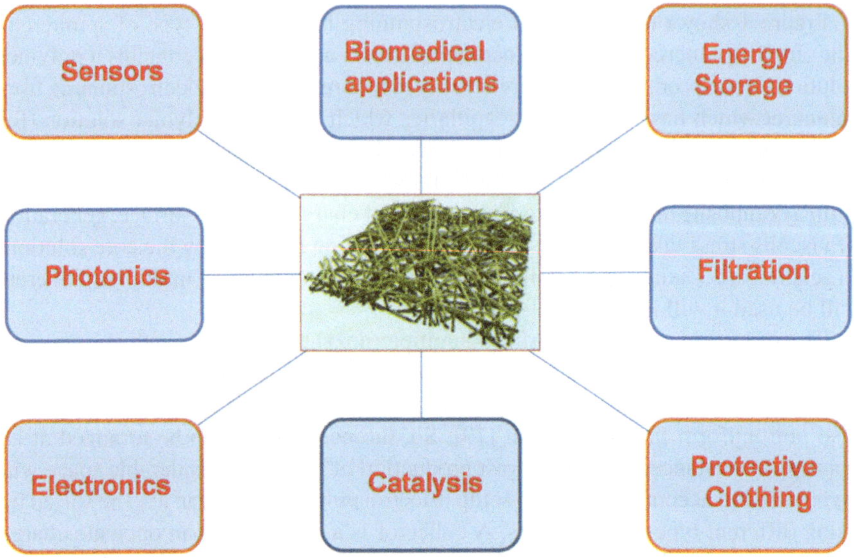

Fig. 2 Application of electrospun nanofibres in different fields

and storage [26, 27], photonic/electronic devices [28], protective clothing [29] and various biomedical applications [30–32] (Fig. 2).

Common precursors used in the electrospinning are organic polymers, small molecules, colloidal particles and composites. Organic polymers like polystyrene (PS) and poly(vinyl chloride) (PVC) for materials for environmental protection [33], PCL, poly(lactic acid) (PLA), and poly(lactic-co-glycolic acid) (PLGA) for biomedical use [34, 35], and poly(vinylidene fluoride) (PVDF) for piezoelectric and/or pyroelectric applications [36, 37], small molecules like amphiphiles (Lecithin) [38, 39] and cyclodextrin derivatives [40] are commonly used for electrospinning. Tetrahydrofuran (THF), dimethyl sulfoxide (DMSO), hexafluoroisopropanol (HFIP), dichloromethane, chloroform, dimethylformamide (DMF), acetone, alcohols and trifluoroethanol are commonly used as solvents in the electrospinning technique.

The research to control the morphology of the nanofibres using electrospinning is gaining more and more attention. Different parameters include the diameter, circular cross-section, and smooth surface of the nanofibres; particularly beaded nanofibres with control over number and size of the bead, core-sheath nanofibres alignment, or hollow nanofibres with controllable hollow cavity diameter and shell thickness, and other unique morphologies.

If we point towards polymer matrices (PM), PM with or without nanoparticles found the application in different fields. Prominently in the biomedical field, for example the use of gum tragacanth-poly (Ɛ-caprolactone)-poly (vinyl alcohol) nanofibrous matrix for diabetic wound healing [41], poly(Lactic Acid) (PLA) membranes as a scaffold in drug delivery of extract of Sedum dendroideum [42], for tissue engineering [43], and also for tissue regeneration [44].

The use of graphene or graphene oxide as fillers gains tremendous attention in the field of material science. Incorporation of graphene or graphene oxides can improve different properties like enhancement in the thermal conductivity of rubber-based silicone composites [45], improved capacitance of the electrodes [46], selective oil absorption [47], dyes absorbtion [47], antistatic properties [48], electromagnetic interference shielding [49], enhanced performance of lithium-ion batteries [50], also mechanical properties [51].

The first part of the chapter deals with the work of dimension control of the polymer matrix using different published examples. Second part briefs about the properties of the nanofibres, which gets improved after the incorporation of graphene or graphene oxide (GO) filler followed by their applications. Incorporation of one-dimensional (1D), two-dimensional (2D) or three-dimensional (3D) nanofillers influences the strength, conductivity, and optical properties of the nanofibres. 1D nanofillers are those nanofillers which possess one of their dimensions less than 100 nm [52, 53]. They exist in the form of nanosheets whose length varies from few nanometers to thousand nanometers, e.g. montmorillonite clay [54], nanographene platelets [55], nanodiscs [56], nanoplates [57], carbon nanowalls [58]. Some examples are layered silicates, MMT [59], ZnO nanodiscs [60], carbon nanowall [61], ZnO nanosheets [62]. Owing to their size and varying shapes, they are useful for the coating, microelectronics, biosensors, and biomedical applications [63].

2D Nanofillers are those nanofillers whose two dimensions are less than 100 nm [52]. They usually exist in the form of filaments, fibres, or tubes. They are the most versatile due to which they found more application as compared to 3D and 1D nanofillers. Examples of 2D nanofillers include carbon nanotubes (CNTs) [64, 65], graphene [65], graphene oxide [66], metal oxides and sulphides [67–69], cellulose and sisal fibres [70, 71]. 2D nanofillers found the application in the field of catalysis [72], energy [73], sensing [74], electronics [75] and optoelectronics [76].

3D nanofillers are the nanofillers whose diameters are less than 100 nm in all the directions. They are also considered as 0D nanofillers, due to their spherical morphology. Other names of 3D nanofillers are nanospheres, nanocrystal or nanogranules. They possess the inherent properties of excellent stability, high refractive index, UV resistance and excellent transparency to visible light and also high photocatalytic activity. Owing to their various properties and low cost, when they mixed with polymer matrix, they are useful for the applications in the field of coatings, separation or purifications, as well as biomedical applications. Examples of 3D nanofillers are quantum dots, carbon black, polyhedral oligomeric silsesquioxane (POSS), nanosilica, semiconductor nanoclusters, etc. [53, 63, 77]. Fan et al. embedded 0D crystalline graphene quantum dots (GQDs) into the activated carbons resulting in the improved electrochemical performance. They utilised D(+)-glucosamine hydrochloride and GQDs and samples of GQD embedded activated carbons (denoted as GEACs) named as pre-GEAC (pre = precarbonized), GEAC, and also without the addition of GQD and named them as pre-AC (precarbonized- activated carbons), AC. They have also prepared another sample, i.e. graphene nanosheet embedded activated carbon (GAC) (Fig. 3). For more details of the preparation, follow the Ref. [78].

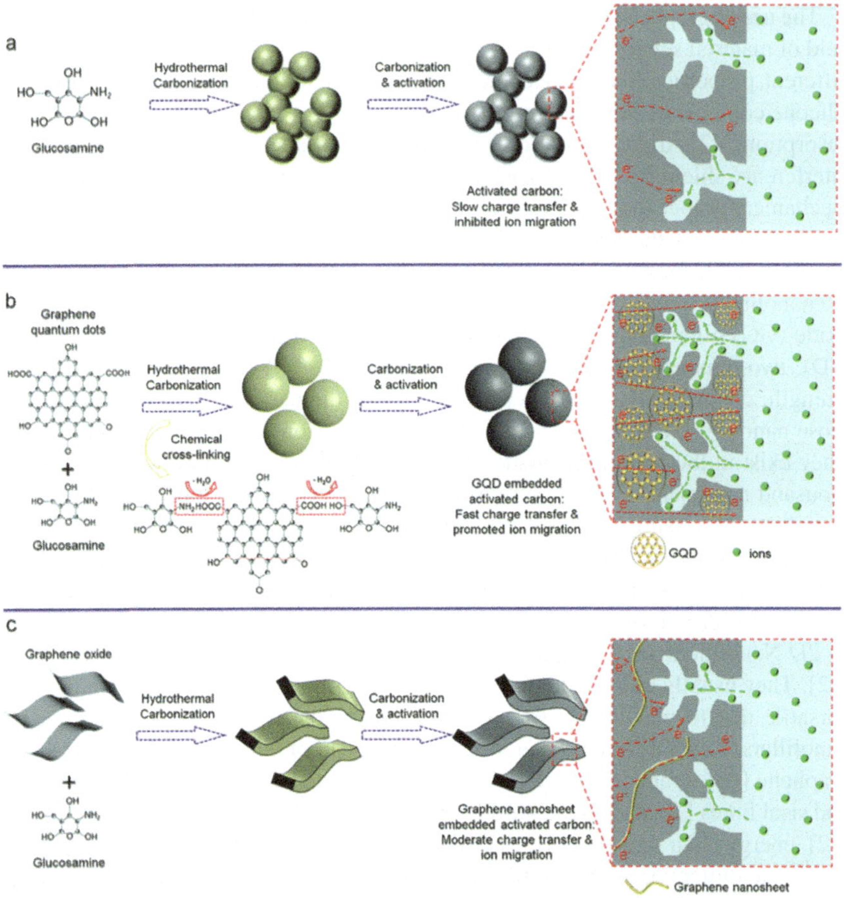

Fig. 3 Schematic representation of the preparation route of the **a** AC, **b** GEAC, and **c** GAC; the GEAC was prepared by hydrothermal carbonisation of the GQD and glucosamine mixture, pre-carbonised followed by the activation in an N_2 atmosphere. The GAC preparation is also the same as that of GEAC, substituting GQD with graphene oxide. Reproduced from Ref. [78] with permission from The Royal Society of Chemistry

They showed that GEAC was much better than pre-AC, GAC, and activated carbon in terms of capacitance and capacitance retention. It showed a capacitance value of 354 FG^{-1} with a retention of 66 and 62% at 50 A g^{-1} and 100 A g^{-1} current densities, respectively. They had also shown only 4% of capacity fading was observed in the case of GEAC in comparison with GAC, which has the capacity fading of 12% when the current density value changed from 50 to 100 A g^{-1}. Thus, they proposed that the ultrafine GQDs were able to form a conductive network of uniform distribution as compared to the large graphene nanosheets in the case of activated carbons, thus promoted the better electrochemical performance particularly at the high current density.

If 0D–2D dimensional materials will be used to incorporate in the polymer matrix that will change the dimension of the polymer matrix, so, achieving a dimension controlled polymer matrix is a challenge. It needs to incorporate a polymer into a network in a space confined to 0D, 1D or 2D space. However, there are reports which selectively made 2D, 3D or fractal dimension polymer matrix.

2 Dimension Controlled 2D Polymer Matrix

Kunitake et al. showed that casted films of bilayer membrane could provide matrices for the production of two-dimensional organic or inorganic molecular networks [79]. They dispersed ammonium amphiphile **1**, bisacrylate **2**, and photoinitiator **3** in water (Fig. 4) and the mixture of **1** and **2** was casted on fluorocarbon filters. The resulting cast films of ammonium amphiphile **1** and its composite with bisacrylate monomer **2** or polymer.

The differential scanning calorimetry (DSC) study showed that the monomer incorporation did not affect the regular bilayer structure of the casted film. Such observation of DSC was due to the microscopic segregation where monomer (or polymer) was accommodated in the inter-bilayer space without dismantling the original bilayer assembly. They established that the obtained highly regular cast film/composite film could act as a better template for the preparation of 2D networks of 20 Å thicknesses. They had also proposed that the method of using cast film was best due to its experimental flexibility, as the dimension of the system can be tuned by changing the amphiphiles or polymers.

Evans et al. explained that the geometrical arrangement of a polymer network could be tailored to obtain the particular macroscopic mechanical property by considering a polymer network having negative Poisson's ratio. Bond hinging, flexure and

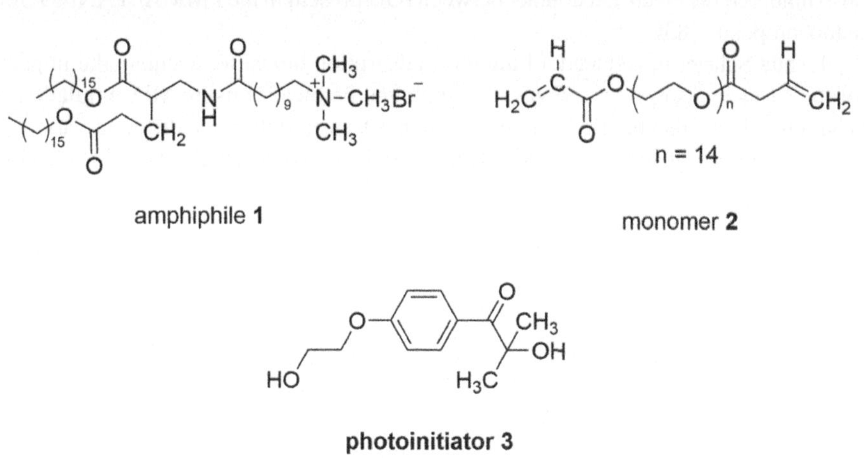

Fig. 4 Structural representation of amphiphile **1**, monomer **2**, and photoinitiator **3**

stretching were the three processes observed for the deformation mechanism of the geometry. They developed a concurrent analytical model to explain networks consisting of extra benzene ring into the vertical branches of the subunits. They found that the predicted stabilised structure was in excellent agreement with the molecular model calculations [80].

McDonald and Koren developed a polymeric composite which can generate nitric oxide. They produced a coating which when reaches to physiological conditions, generates nitric oxide, and the generated nitric oxide can inhibit the platelets aggregation. The embodiments which produced the nitric oxide contained nitrocellulose or other functional groups which possess $N_2O_2^-$ group attached to amino groups of the polymer [81].

Carvalho et al. prepared a 2D coordination polymer matrix, ∞ [Gd(DPA)(HDPA)] for solid-phase extraction, by reacting a mixture of pyridine-2,6-dicarboxylic acid and gadolinium nitrate in the water at a high temperature of 168 °C. They proposed that these polymers can be used for the analysis of pesticides in the medicinal herbs, particularly phosalone, tetradifon and bifenthrin compounds present in the herb, Cordia salicifolia. This property makes these polymers useful for the screening of pesticides present in herbs by pharmaceutical of industries and laboratories [82].

Huang et al. prepared 2D hybrid sheets using polyaniline decorated reduced graphene oxide (rPANI@rGO). They carried out two steps for the preparation of these sheets, i.e. (i) in situ polymerisations of aniline on graphene oxide (ii) reduction of the sheets obtained in the step (i) using hydrazine (Fig. 5).

They utilised rGO and rPANI@rGO as a filler for poly(methyl methacrylate) (PMMA) nanocomposites and proposed that PMMA/rPANI@rGO nanocomposites showed a high dielectric constant with a low dielectric loss. So the property can be tuned by varying the amount of hybrid sheets in PMMA/rPANI@rGO nanocomposite. They proposed that the improvement in the dielectric properties was due to isolation effect of rPANI on rGO, which in turn improved the GO dispersion and also hindered the electrical contact between rGO present in the PMMA/rPANI@rGO nanocomposite [83].

Joshua Schrier introduced a Langmuir-adsorption model to calculate the impact of the surface adsorption of a gas on membrane permeance using MD simulations. That model can also be useful to anticipate gas transportation via different varieties of the 2 D porous materials [84, 85]. Transport of molecules can also be controlled by

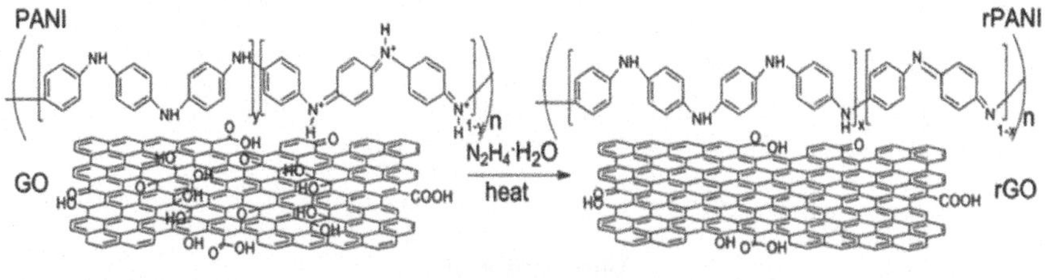

Fig. 5 Schematic presentation of GO and PANI reduction using hydrazine. Reproduced from Ref. [83] with permission from The Royal Society of Chemistry

tuning the surface adsorption, e.g. Fluorination of the surface can reduce the attractive dispersion interactions, by utilising a non-uniform distribution of charges facilitating the adsorption of the dipolar molecules [86]. He also proposed that a similar effect based on the different wetting of the surface can be useful to separate the mixture of the liquids, i.e. by tuning the surface by hydrophilic or hydrophobic surface modification, and also the distribution of the charges like interaction with polar molecules or ions inside the nanoporous sheets. He utilised molecular dynamics (MD) simulations to characterise 2D hydrocarbon polymer porous Graphene-E-Stilbene1 (PG-ES1). The PG-ES1polymer was used to separate carbon dioxide gas from methane, nitrogen and oxygen gases with a permeance of 3×10^5 gas permeation units (GPU). He also proposed that PG-ES1can become the first economical material to meet the demand of post-combustion CO_2 capturing from fossil fuel industries and other applications [87].

3 Dimension Controlled 3D Polymer Matrix

In 1988, Alaeddinoglu et al. developed a simple acrylate gel to form 3D polymer matrix by incubating acrylamide, N,N'-methylene bisacrylamide and ammonium persulphate for 90 min at 37 °C [88] and immobilise it with yeast. They proposed that if a concentrated polymer solution or higher cross-linked polymer is made using a polymer solution with high concentration, then it will have a low swelling ratio. They used the yeast for the invertase enzyme and found that if the water content of the matrix will be high, the activity of the invertase enzyme will also be high. Alaeddinoglu's group also proposed that if the cell content of the gel will be high, the enzyme activity will be low.

Mooney et al. improved the porosity with well-defined structure and pore size. They used a solid porogen NaCl and partly fused it to polymer matrix in 95% humidity before the formation of the continuous polymer matrix. The fabrication of three-dimensional porous scaffolds of the copolymer lactide/glycolide with a ratio of 85:15 was carried out by either a solvent casting/particulate leaching process or a gas foaming/particulate leaching process. They proposed that 24 h of salt crystal fusing increased their size, or radius of curvature, which increase a two-fold ability to compress in case of solvent cast scaffolds. In contrast, there was a decrease in compress ability in the case of the use of gas foaming. The total porosity for solvent cast scaffolds and gas foaming scaffolds were 97 ± 1%, and 94 ± 1%, respectively. Such a highly connected network of scaffolds is useful for enhanced cell migration, more number of cellular interactions, and improved vascular and neural growth into tissue-engineered scaffolds [89].

Hydrogel synthesis is one of the best examples in 3D dimension control. Sahiner showed that cross-linked hydrogel of poly (2-acrylamido-2-methyl-1-propansulfonic acid) [90] of approximately similar structure was prepared on UV irradiation of AMPS monomer' aqueous solution of containing different

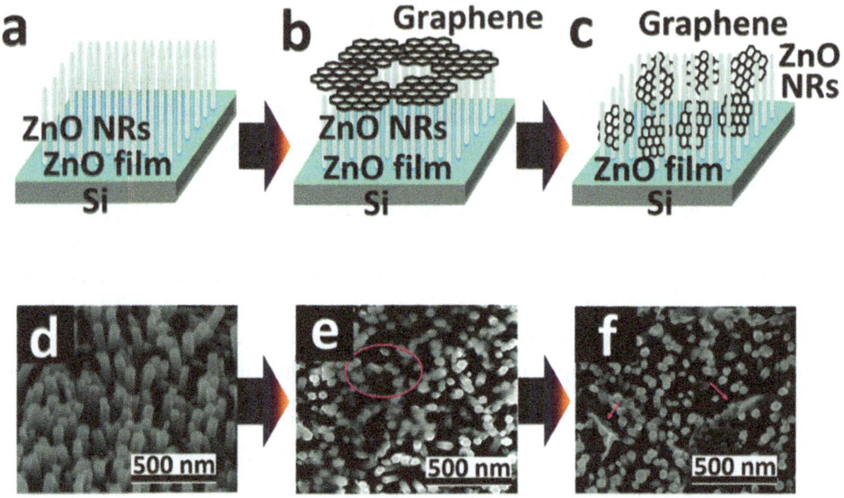

Fig. 6 Schematic representation where **a–c** are depicting the formation of ZnO NRAs–graphene hybrid nanoarchitectures. FESEM images of **d** ZnO nanorod arrays (NRAs) **e** top view image of graphene layers on the surface of the NRAs and **f** final ZnO–graphene hybrid nanoarchitectures. Reproduced from Ref. [76] with permission from The Royal Society of Chemistry

amounts of N,N′-methylenebisacrylamide. Three different percentage of N,N′-methylenebisacrylamide (1, 0.75, and 0.5%) were used. Morphologies did not show much change, but the swelling ability of the hydrogel was significantly affected. Also, other important properties like metal-ion loading capacity and metal-ion formation [91] were also significantly affected [90].

Qurashi et al. synthesised nanofiller graphene–ZnO hybrid nanoarchitecture by the sonochemical process. As shown in Fig. 6, the steps involved (a) ZnO film was grown on the silicon substrate, (b) then, the ZnO nanorods were grown on the ZnO film, (c) the graphene sheets were transferred on the ZnO nanorod vertical arrays top surface followed by the intercalation of the graphene sheets between the ZnO NR arrays by sonochemical treatment.

They proposed that graphene surface and interface with ZnO were working as an anchor for the charge carriers, i.e. electrons and holes. Ambipolar graphene was also suitable for instant transfer of both electrons and holes, thus can make an excellent hybrid electronic material. And they found the property exciton formation by graphene and ZnO hybrid. Therefore, they conducted optoelectronic experiments. They had finally proposed that more graphene loading could improve the reflex action of the hybrid nanomaterial devices upon optical excitation.

Hydrogels can also entrap hydrophobic drugs because of their multiple polymer chains [92]. Hence, if the parameters like pH, temperature, or enzymatic action are well understood; the dimension of the hydrogel can be controlled, which governs the release of the drug molecule at a particular location [93]. Other properties like biocompatibility, biodegradability, and their chemical structures also significantly affects their hydrophilic and mechanical properties [94]. For example, Zheng et al. embedded upconverting nanoparticles (UCNPs) into a hydrogel which possess a substituent that can be activated by light. They proposed a strategy of controlled activation of angiogenesis and 3D cell spreading using near-infrared (NIR). They suggested that UCNPs could convert NIR to UV emission resulting in the photochemical reaction. They showed that human vascular endothelial cells (HUVEC) when embedded in their hydrogels could form vascular networks geometries which were governed by the irradiation pattern. They utilised PEG thiol/maleimide hydrogels [95] containing UV sensitive/triggerable cell adhesive peptide (cyclo [RGD(DMNPB)fC]) [96, 97], and embedded these hydrogels with UCNPs coated with poly(maleic anhydride-alt-1-octadecene) (PMAO), along with a vascular endothelial growth factor (VEGF), and HUVEC, and perform their study.

Figure 7 shows that in the presence of NIR radiation, UCNPs produces UV emission, as cyclo [RGD(DMNPB)fC] is UV triggerable; it releases a byproduct due to the photocleavage. Such photoactivation leads to the angiogenesis. They did not detect any cell damage. Thus, they regulated angiogenesis using NIR in a hydrogel.

Another example is a disulphide-cross-linked (poly(oligo(ethylene-oxide) monomethyl ether methacrylate)) (POEOMA) nanogel which gets degrade specifically by the reducing action of glutathione making it suitable for targeted drug delivery application [99]. Similarly, another example includes a biodegradable

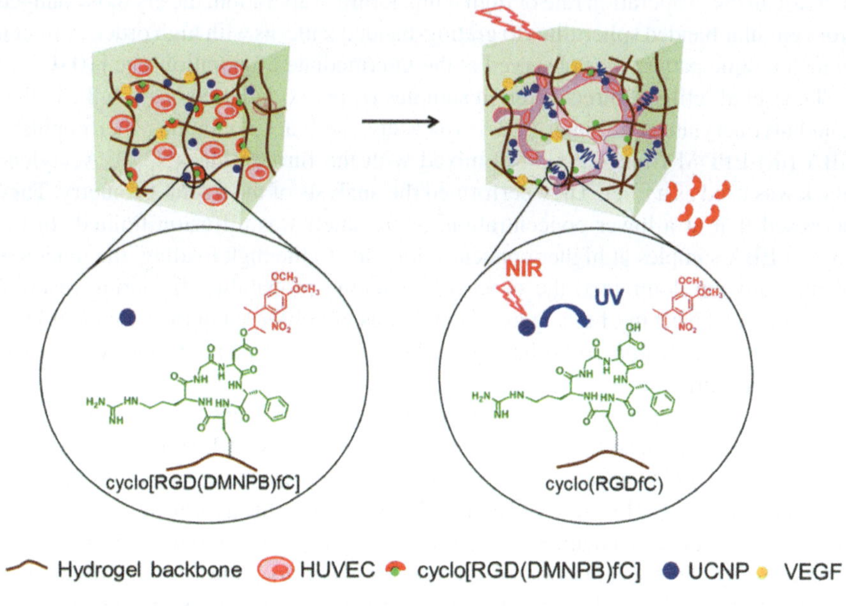

Fig. 7 Pictorial representation of NIR light-induced angiogenesis in a photoactivatable hydrogel embedded with UCNP-PMAOs. Reproduced from Ref. [98] with permission from The Royal Society of Chemistry

polyurethane family, Novosorb, which was also utilised for the generation of a 3D dermal matrix [100].

4 Dimension Controlled Fractal Dimension Polymer Matrix

Mandelbrot gave the geometrical concept of fractal dimension which deals with the system with stochastic self-similarity, and non-integer dimensionality [101]. Amita Chandra proposed that about anion clustering and fractal pattern growth in ion-conducting polymeric matrix of poly(ethylene-oxide) (PEO):NaI or (PEO):NH_4I polymer by considering the example of mobile iodine anion I_3^-. It was proposed that iodine forms the fractal pattern in the ion-conducting polymeric matrix. In contrast, a tree-like anisotropic pattern was created at the interface [102]. In other words, fractal growth pattern means a dendrimer like growth pattern.

Woo et al. crystallised phthalic acid in the presence of tannic acid (TA) to find out its growth dimension. They investigated the crystal assembly of phthalic acid and found out the relation between banded patterns and branching structure. In other terms, they found the way to develop a micrometre thin film of the phthalic acid in the fractal dimension. After utilising the different ratio of ethanol/water solution and different temperature to grow the phthalic acid crystal, they proposed that with the increase in the evaporation rate or high temperature evaporation, the crystals changed from circular banded spherulites to grating-banded patterns with high order. A fractal branch unique pattern was observed at the intermediate evaporation rate [103].

Jäger et al. utilised three different samples (i) peroxide-based cross-linked ethylene butyl acrylate copolymer (EBA) (ii) acetylene black containing thermoplastic EBA (iii) EPDM Buna EP G5440 mixed with the furnace blackN330. Acetylene black was used as a filler. They performed the analysis of the fractal geometry. They proposed that at a lower concentration, aggregation was diffusion-limited. In the case of EBA samples at higher concentration, due to the high loading, the tendency of the collision dominated the process of sticking probability, favouring reaction controlled mechanism of the aggregation. It was not observed in the case of EPDM, so it remained diffusion-controlled; thus, the structure can be interpreted even at the higher concentration.

They showed that when the concentration of the carbon black increased in the case of EBA-based samples, there is a crossover happens from diffusion-limited cluster aggregation (DLCA) mechanism to reaction limited cluster aggregation (RLCA) mechanism. Due to RLCA, in the case of EBA samples, the aggregates number was not concentration-dependent. Still, the fractal dimensions were found to be increased on increasing the concentration of the carbon black resulting in the formation of smaller clusters. Thus, the fractal dimension was concentration-dependent. Same observations of concentration dependency of the fractal dimension were observed in the case of EPDM sample. It was found that the fractal dimension of EBA was

1.8 at a lower concentration of acetylene black, which rises to 2.1 on increasing its concentration. They had also observed a similar trend in the case of EPDM Buna EPG 5450 on the addition of furnace black N330 as filler; the dimension changed from 1.9 to 2.4. Later, they tried to establish a connection between electrical conductivity and fractal geometry, but no connection could be made [104].

Kozlov and Lipatov also utilised two different composites made up of poly(hydroxyether) and graphite with different filler concentration and investigated their adhesion and interfacial layers. They proposed that characteristics like interfacial layer, adhesion and mechanical properties are interconnected with the difference between polymer matrix and fractal dimensions of the surface of the aggregates of filler particles. In other words, the structures of the polymer matrix were distorted due to the different filler surface [105]. They also had determined these distortions or disturbances by utilising the change in the fractal dimensions (D_f) of polymer matrix structures [106]. Mousavi et al. who studied the surface morphology of PANi/G composites using SEM and the fractal dimension concept. They proposed that the cyclic voltammetry (CV) and electrochemical impedance spectroscopy (EIS) are the best methods to find the most reliable value of the fractal dimension (D_f) [107].

5 Graphene Filler-Based Nanofibres by Electrospinning

Electrospin techniques can control parameters like the molecular weight of the polymer along with its architecture. Variables like the electric field, capillary-collector distance, and flow rate of the polymer solution easily in the electrospinning technique can alter the properties of the solution like conductivity, the viscosity of the solution [108]. Electrospinning can also tailor the electrospun nanofibre polymer matrix by adjusting the component ratio, and also by using different additives like GNC and CNTs [109]. Nanofibres having different surface morphologies, cross-sectional shapers, branching, coil, zigzags, and the bead can also be synthesised [110].

Herein, an interesting example of by Chronakis et al. which shows the effect of solvents on the morphology and fibre diameter on the phospholipid electrospun nanofibres [39]. They utilised asolectin solution for their study using different solvents, i.e. isooctane, cyclohexane and limonene and compared them with $CHCl_3$: DMF (3:2 v:v) solvent solution. It was found that all solvents were capable of producing electrospun phospholipid nanofibres of asolectin. At the lower/optimised concentration, fibrous morphology was observed in each solvent without any beads. However, when the concentration of asolectin was kept 45% w/w, different morphologies were observed in different solvents. Fibrous morphology was observed in the case of $CHCl_3$:DMF, whereas beaded short fibres were seen in cyclohexane, and beaded morphology was seen in the case of both limonene and isooctane (Fig. 8). They proposed that the electrospinnability of various biomaterials such as cyclodextrins, polypeptides, supramolecular molecules could be improved using different solvents.

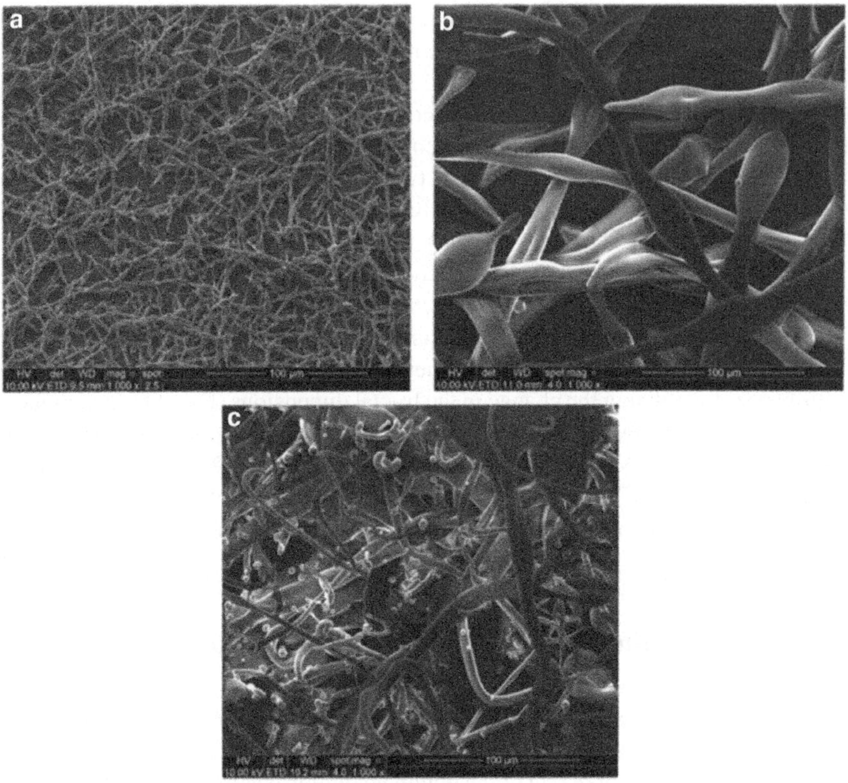

Fig. 8 Representation of the SEM images of the morphology of the electrospun asolectin solutions using a single needle **a** 40% w/w asolectin in $CHCl_3$/DMF [3:2 v/v], scale bar:100 μm, **b** 45% w/w asolectin in cyclohexane, scale bar:100 μm **c** 55% w/w asolectin in isooctane, scale bar:100 μm. Reproduced from Ref. [39] with permission from The Royal Society of Chemistry

They had also studied the co-axial electrospinning of the asolectin using different solvents in inner and outer needle. It was observed when they utilised $CHCl_3$: DMF (3:2 v:v) in the inner needle and DMF, $CHCl_3$, isooctane, and cyclohexane in the outer needle, the value of the average diameter of the nanofibres were 0.38 μm, 0.66 μm, 1.01 μm, 1.54 μm, respectively using different parameters like feed rate, the distance between jet and collector, and the voltage, at the concentration of 4 w/w. They proposed that the high dielectric constant of the DMF solvent had a significant effect on the morphology of the nanofibres.

They also showed that the diameter of the asolectin fibre was also correlated with the dielectric constant and boiling point of the solvent. The viscoelastic property of the solution dominates at a certain point, where electrostatic force cannot further make the jet thin, which in turn affects the diameter of the electrospun fibres. They proposed that asolectin solutions in isooctane, cyclohexane and limonene showed onset of gelation near their concentration at which the nanofibres could be produced

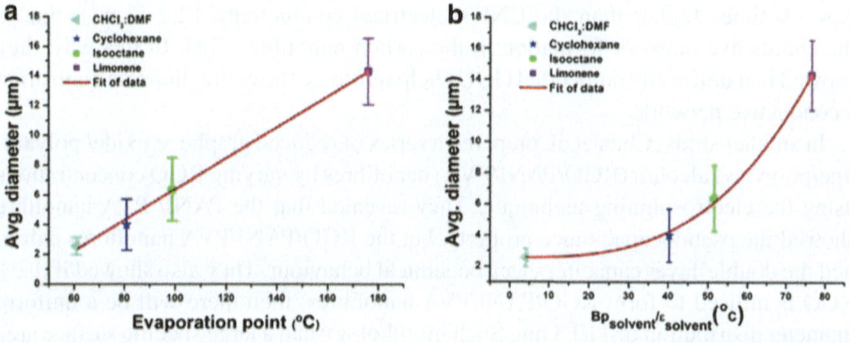

Fig. 9 Representation of the average diameter of electrospun phospholipid fibres as a function of **a** solvent evaporation rate and **b** boiling point and the dielectric constant of the solvents. Reproduced from Ref. [39] with permission from The Royal Society of Chemistry

(62% w/w), but when 45% v/v of asolectin in $CHCl_3$/DMF solvent was used, no gelation was observed even after using the 65% w/w ratio.

They showed the linear relationship between the diameters of the nanofibres with the evaporation rate of the solvent. The diameter of the nanofibres was highest in the case of limonene as compared to the other solvents used for the electrospinning of the asolectin. They had also found a similar trend in the diameter, after plotting the curve between the ratio of boiling point and dielectric constant (Fig. 9).

Mainly, graphene-based fillers are useful to produce high-performance material for energy storage [111], tissue engineering [112, 113], water treatment [114] and photocatalysis [115] due to which they are gaining the utmost attention in today's scenario. As graphene or graphene oxide filler-based polymer composites are highly conductive, they can also be used in applications like electrostatic discharge devices, electrical switches, ambipolar field-effect transistors, sensors, electrode and electromagnetic interference (EMI)-shielding materials [116]. In detail, we shall discuss the properties which get improved after the incorporation of graphene fillers and their applications in the following section.

Recently Fan et al. utilised 0D graphene quantum dots (GQDs) to improve the surface area, conductivity and mechanical strength of the carbon nanofibres fabrics (CNF) [117]. They electrospun GQD into CNF to form a GQD-reinforced Electrospun Carbon Nanofibres Fabrics (GRCNF) which possessed high surface area, improved conductivity and high mechanical strength. They also had electrospun polyacrylonitrile powder (PAN) and GQDs using DMF solution and oxidised the electrospun PAN to obtain PO-G/PAN, whose carbonisation gave GRCNF. They have also prepared activated GRCNF by the chemical activation of PO-G/PAN. It was noticed that GRCNF surface area reduced to 44 m^2g^{-1} as compared to the pure CNFs, 140 m^2g^{-1}, due to high density of inner structure and large diameter of the fibre. Whereas in the case of AGRCNF, the surface area was found to be 2032 m^2g^{-1} with a developed porous structure which had 2.5 times higher strength than CNF. They noticed that the electric conductivity of the GRCNF was 96.4 S m^{-1} which was 7.9 times higher than the CNF's electrical conductivity 12.2 S m^{-1}, due to the conductive network formation in the carbon nanofibres [78]. In this way, they showed that uniformly embedded GQDs help reinforce the entire phase and construct a conductive network.

In another study, Chen et al. prepared a series of reduced graphene oxide/ polyaniline/polyvinyl alcohol (RGO/PANI/PVA) nanofibres by varying RGO concentrations using the electrospinning technique. They revealed that the PANI/ PVA nanofibre showed the pseudocapacitance property, but the RGO/PANI/PVA nanofibres exhibited the double-layer capacitor electrochemical behaviour. They also showed if 0.2% RGO is utilised to form RGO/PANI/PVA nanofibres, then there will be a uniform diameter distribution of 119.8 nm. Such morphology had a large specific surface area of fibres and provided enough channels to transport ions. RGO/PANI/PVA nanofibres showed higher specific capacitance of 174 F g^{-1} as compared to PANI/PVA with a specific capacitance of 105 F g^{-1}, making RGO/PANI/PVA nanofibres as a suitable candidate for supercapacitor application [118].

Mohamed et al. prepared polyacrylonitrile (PAN) nanofibres whose mechanical properties were reinforced by the utilisation of graphene oxide. Fabrication of PAN/GO nanofibres was performed using electrospinning. Their tensile tests showed that the tensile strength and Young's modulus was improved by 64.6 and 71.4%, respectively, on the addition of 1% GO into the PAN nanofibres. They proposed that the inclusion of GO fillers enhanced the mechanical strength of the PAN nanofibres [119]. They found on increasing the GO percentage to 1.5%, the value of tensile strength, Young's modulus and elongation at break were significantly reduced. Thus, they proposed that if the percentage of the GO filler was increased to 1.5%, the resulting fibres would be more brittle and fragile due to the agglomeration of the GO [120, 121].

Correa et al. developed polycaprolactone (PCL) nanofibre [122] mats embedded with photoresponsive nanogel containing AgNps for biomedical applications. Electrospinning of PCL and AgNps was done in 3:7 ratio using DCM and DMF as solvents [123]. They proposed that the immobilisation of Ag NPs on the surface of the nanofibres took place by H-bonds and electrostatic interactions between amine functional groups of nanogels and negative charge of nanofibres. It was found that developed smart mats were useful in wound dressing [124], that can be activated by light, and also control the bacterial infections from Gram-positive *Staphylococcus aureus* and Gram-negative *Escherichia coli.* They proposed when 405 nm light was used, the plasmonic band of silver nanoparticles breaks the nanogel, due to which the AgNPs get released. In other words, the propagation of silver ions was controlled by light [125].

Asadi et al. utilised a plant protein zein, and electrospun it with GO composite to form a nanofibrous dressing [126]. They found that the GO was uniformly distributed and blended in the polymer matrix. Zein nanofibres usually suffer from less mechanical stability for dressing material. However, the incorporation of GO increased its mechanical strength with the increase in hydrophilicity, which provided a suitable condition for proliferation and cell attachment. Also, GO acts as a nanocarrier of tetracycline hydrochloride [127], which provide more controlled drug release. And, due to nanocarrier's bactericidal ability [128] and non-toxic nature [129, 130], wound healing gets faster. They proposed that zein-GO nanofibres are excellent for the use of wound dressing due to its biocompatibility and mechanical properties.

Wound healing, especially in the case of the glycemia, is always attracting attention in the medical field. Different studies showed that the delayed wound healing in the case of diabetic ulcers was due to the destabilisation of the hypoxia-inducible factor 1 a (HIF-1a) [131, 132]. Thus, it is important to design HIF-1a stabilising materials for the wound dressings to resist the hyperglycemic effect. Dimethyloxalylglycine (DMOG) is a well-known competitive inhibitor of prolyl hydroxylases, which can stabilise as well as activate HIF-1a [133]. So, Chen et al. developed DMOG delivering nanofibrous wound dressings for wound healing in the case of diabetes [134]. They designed three samples (i) co-axial nanofibres loaded with DMOG in the core layer (C-DPC) (ii) Mono-axial poly(e-caprolactone) (PCL)/type I collagen(Col I) electrospun nanofibres with DMOG (M-DPC) (iii) Mono-axial poly(e-caprolactone) (PCL)/type I collagen(Col I) electrospun nanofibres without DMOG (PC). When the concentration of the DMOG was kept at 0.33 mg in both the samples, C-DPC, M-DPC. Maximum drug release was observed in the case of M-DPC, whereas the cell proliferation and migration of fibroblasts on different nanofibre-based wound dressing was highest in the case of C-DPC. HIF-1a level was also observed in the case of C-DPC.

They treated the diabetic wound with and without nanofibrous dressings which are shown in Fig. 10 on days 0, 3, 7, 14 days. It was observed that the wounds, when covered with the C-DPC, showed faster healing in the entire wound healing process. They proposed that on an average of up to 48% of the wound was contracted on the 3rd day. Whereas 66% of healing rate was observed in the case C-DPC followed by M-DPC and PC with values of 59% and 57%, respectively after 7th day. After 14 days, the wound in all the cases was almost healed, but the maximum wound healing of 99.8% was observed in the case of C-DPC case (Fig. 10). Thus, their results proposed that DMOG delivering nanofibres may be used as a promising functional wound dressing material for the healing of diabetic wounds.

Jo and Erdenebileg et al. developed a nanofibrous polyurethane-nanographene oxide (PU-nGO) scaffolds resembling a natural extracellular matrix (ECM). When incorporated with GO, the improvement of hydrophilicity, elasticity, and stress relaxation capacity of PU was observed. They proposed that the PU-nGO nanofibrous membrane not only improved the initial adhesion but also increased proliferation of C2C12 cells with little cytotoxicity. Also, their nanofibres increase myogenic mRNA and myosin heavy chain (MHC) expression level and increase myotubular production. In such a manner, they suggested that PU-nGO nanofibres can serve as a potential matrix for the engineering of skeletal muscles [135].

Simzar and Hossein et al. fabricated a bone architecture ECM to study their 3D scaffolds having PTFE and PVA, with and without graphene oxide (GO) nanoparticles. They proposed that these 3D scaffolds can act as an ideal candidate for bone tissue engineering as the 3D templates [136].

Nagalakhsmi et al. reported the nonlinear absorbing behaviour of *m*-nitroaniline (mNA) organic molecules incorporated in the core of a Poly (methyl methacrylate)

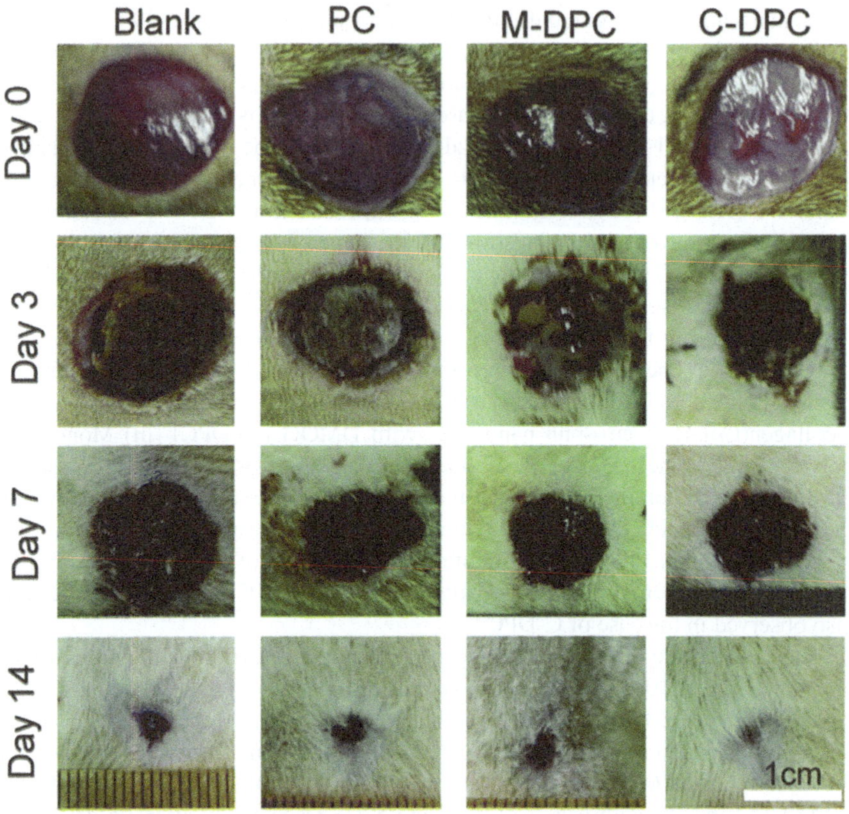

Fig. 10 Pictorial presentation of the diabetic wounds treated with blank, PC, M-DPC, C-DPC nanofibrous wound dressings. Reproduced from Ref. [134] with permission from The Royal Society of Chemistry

(PMMA) and Poly (ethylene-oxide) (PEO) fibre using electrospinning to form free-standing films. They observed the nonlinear switching behaviour of the prepared films in mNA-PMMA nanofibre from reverse saturable absorption (RSA) to saturable absorption (SA). This transition process took place via a W-pattern for the increase in pulse energy. In contrast, the mNA–PEO polymer matrix showed intensity-dependent SA behaviour. They proposed that such nonlinear absorption is of interest for different optical applications like optical sensors, switching and so on. They demonstrated that mNA-PMMA fibre matrix could be employed to make the laser safety devices. They have also shown that fibre matrix free-standing films are of significant interest for the laser photonic devices formation [137].

Amirkianoosh et al. studied the optical and electrical effect of using 2D graphene flakes into electrospun PCL fibres. They noticed that the electrical conductivity gets increased on increasing the concentration of graphene into PCL fibres, similar trend was observed for the light absorption capability of the PCL fibres which was confirmed by specular and diffuse reflectance spectroscopy, so PCL fibres can be utilised to make optical filters for controlled light absorption/reflection [138].

Gao et al. synthesised thermoplastic polyurethane (TPU) nanofibres by electrospinning and later used ultra-sonication to decorate graphene onto it. They then modified the obtained composite with polydopamine (PDA) followed the hydrophobic treatment using 1H, 1H, 2H, 2H perfluorodecanethiol (PFDT). It was found that PDA introduction improved the interaction between the TPU and graphene, and also among graphene sheets itself. They proposed that the tensile strength and Young's modulus was enhanced on the addition of graphene sheets as compared to TPU nanofibrous membranes. These conductive polymer nanofibre composites (CPNCs) displayed excellent cyclical stability, controllable sensitivity, and high stretchability when used as a sensor to detect strain. So, Gao et al. proposed that exact motions of human can be monitored using these strain sensor on the skin; thus, these CPNC can be a promising candidate for making wearable devices [139].

Kun Dai et al. synthesised a multifunctional strain sensor by the electrospinning process, followed by ultra-sonication and *in situ* polymerisation of dopamine. They have also used the TPU fibre incorporated with RGO. They embedded the polydopamine (PDA) on the surface of RGO/TPU fibres. The PDA/RGO/TPU (PRT) fibre mats work as a strain sensor which exhibited excellent performance with high stretchability, durability, higher breathability and showed more sensitivity and faster response time. Their PRT fibrous mats showed a high sensitivity for humidity and organic gases, and other environmental stimuli [140].

Hayat et al. synthesised a GO-PVA composite. They modified screen-printed carbon electrodes (SPCE) through electrospinning of GO-PVA composite. They showed that very fine nanofibres were covering the SPCE. Heavy metal detection in water samples was performed by using the produced sensitive electrochemical enzyme sensor. They further immobilised alkaline phosphate on the surface of GO surface and used them for the sensing of heavy metals, i.e. lead (Pb^{+2}), mercury (Hg^{+2}) and cadmium (Cd^{+2}) cations. They utilised a unique combination of alkaline phosphate (ALP) on GO and heavy metal sensing and proposed that this combination showed good ability to sense the toxic heavy metal cations. Their ALP bases sensor was specific for the detection of Hg^{+2} among all three metal cations for the inhibition. Since SPCEs are not costly, they can be used to make a single-use biosensor. Reproducibility and precision were also checked, which showed an RSD value of 3.25. They suggested that the covalent interaction between the 4-carboxyphenyl modified electrode and primary amine functional group in the biomolecules could restrict the ALP leaking, which in turn provided the electrode with extra stability [141]. Various other biosensors are also made using graphene or graphene oxide as filler like carbon-modified nanofibre biosensor for immobilisation of RBL mast cells [142], 17α-ethynyl estradiol (EE2) detection [143], and detections of glucose [144].

6 Conclusion

The chapter successfully showed different examples showing the methods to control the 2D, 3D and fractal dimensions of the polymer matrix. With a brief of electrospinning, it also described the improved properties of different materials with the utilisation of graphene as filler. In the end, it gives examples in various fields like the biomedical field, optics, sensors and more.

Acknowledgements AS would like to acknowledge the Ministry of Human Resource Development (MHRD), the Council of Scientific Industrial and Research (CSIR) and IIT Indore. SM would like to acknowledge the Science and Engineering Research Board (SERB) for the National Post-Doctoral Fellowship (NPDF). SKS would like to thank the CSIR for the Senior Research Fellowship.

References

1. Gilbert, W.: De magnete, magneticisque corporibus, et de magno magnete tellure; Physiologia nova, plurimis and argumentis, and experimentis demonstrate (1600)
2. Nollet: Lectures in Experimental Philosophy. Printed for S. Austen, London (1748)
3. Rayleigh, L.: XX. On the equilibrium of liquid conducting masses charged with electricity. London Edinburgh Dublin Philos. Mag. J. Sci. **14**, 184–186 (1882)
4. Electrostatic Spraying of Liquids. Von A. G. Bailey, Research Studies Press LTD Taunton, Somerset/John Wiley and Sons Inc, New York 1988, 197 Seiten. Physik in unserer Zeit. **20**, 160–160 (1989)
5. Qin, X.: 3—co-axial electrospinning of nanofibers. In: Afshari, M. (ed.) Electrospun Nanofibers. Woodhead Publishing, pp. 41–71 (2017)
6. Zhang, C., Feng, F., Zhang, H.: Emulsion electrospinning: fundamentals, food applications and prospects. Trends Food Sci. Technol. **80**, 175–186 (2018)
7. Lian, H., Meng, Z.: Melt electrospinning vs. solution electrospinning: a comparative study of drug-loaded poly (ε-caprolactone) fibers. Mater. Sci. Eng. C. **74**, 117–123 (2017)
8. Díaz, J.E., Fernández-Nieves, A., Barrero, A., Márquez, M., Loscertales, I.G.: Fabrication of structured micro and nanofibers by coaxial electrospinning. J. Phys. Conf. Ser. **127**, 012008 (2008)
9. Zhang, L.-H., Duan, X.-P., Yan, X., Yu, M., Ning, X., Zhao, Y., Long, Y.-Z.: Recent advances in melt electrospinning. RSC Adv. **6**, 53400–53414 (2016)
10. Das, S., Wajid, A.S., Bhattacharia, S.K., Wilting, M.D., Rivero, I.V., Green, M.J.: Electrospinning of polymer nanofibers loaded with noncovalently functionalised graphene. J. Appl. Polym. Sci. **128**, 4040–4046 (2013)
11. Ramakrishna, S., Fujihara, K., Teo, W.-E., Lim, T.-C., Ma, Z.: An Introduction to Electrospinning and Nanofibers, p. 396. World Scientific (2005)
12. Alghoraibi, I., Alomari, S.: Different methods for nanofiber design and fabrication. In: Barhoum, A., Bechelany, M., Makhlouf, A. (eds.) Handbook of Nanofibers, pp 1–46. Springer International Publishing, Cham (2018)
13. Goldstein, A.S., Thayer, P.S.: Chapter 11—Fabrication of complex biomaterial scaffolds for soft tissue engineering by electrospinning. In: Grumezescu, A.M. (ed.) Nanobiomaterials in Soft Tissue Engineering, pp 299–330. William Andrew Publishing (2016)
14. Feng, X., Li, J., Zhang, X., Liu, T., Ding, J., Chen, X.: Electrospun polymer micro/nanofibers as pharmaceutical repositories for healthcare. J. Control. Release **302**, 19–41 (2019)

15. Asmatulu, R., Khan, W. S.: Chapter 2—Historical background of the electrospinning process. In: Asmatulu, R., Khan, W.S. (eds.) Synthesis and Applications of Electrospun Nanofibers, pp 17–39. Elsevier (2019)
16. Long, Y.-Z., Yan, X., Wang, X.-X., Zhang, J., Yu, M.: Chapter 2—Electrospinning: the set-up and procedure. In: Ding, B., Wang, X., Yu, J. (eds.) Electrospinning: Nanofabrication and Applications, pp 21–52. William Andrew Publishing (2019)
17. Rutledge, G.C., Fridrikh, S.V.: Formation of fibers by electrospinning. Adv. Drug Deliv. Rev. **59**, 1384–1391 (2007)
18. Aravindan, V., Sundaramurthy, J., Suresh Kumar, P., Lee, Y.-S., Ramakrishna, S., Madhavi, S.: Electrospun nanofibers: a prospective electro-active material for constructing high performance Li-ion batteries. Chem. Commun. **51**, 2225–2234 (2015)
19. Migliaresi, C., Ruffo, G. A., Volpato, F. Z., Zeni, D.: Advanced electrospinning setups and special fiber and mesh morphologies (2012)
20. Huang, Z.-M., Zhang, Y.Z., Kotaki, M., Ramakrishna, S.: A review on polymer nanofibers by electrospinning and their applications in nanocomposites. Compos. Sci. Technol. **63**, 2223–2253 (2003)
21. MartinIan, G.E., Cockshott, D., Fildes, F.J.T.: US4044404A (1977)
22. Gopal, R., Kaur, S., Ma, Z., Chan, C., Ramakrishna, S., Matsuura, T.: Electrospun nanofibrous filtration membrane. J. Membr. Sci. **281**, 581–586 (2006)
23. Hemalatha, R.G., Ganayee, M.A., Pradeep, T.: Electrospun nanofiber mats as "smart surfaces" for desorption electrospray ionization mass spectrometry (DESI MS)-based analysis and imprint imaging. Anal. Chem. **88**, 5710–5717 (2016)
24. Lu, P., Xia, Y.: Novel nanostructures of rutile fabricated by templating against yarns of polystyrene nanofibrils and their catalytic applications. ACS Appl. Mater. Interfaces **5**, 6391–6399 (2013)
25. Dai, Y., Lim, B., Yang, Y., Cobley, C.M., Li, W., Cho, E.C., Grayson, B., Fanson, P.T., Campbell, C.T., Sun, Y., Xia, Y.: A sinter-resistant catalytic system based on platinum nanoparticles supported on TiO_2 nanofibers and covered by porous silica. Angew. Chem. Int. Ed. **49**, 8165–8168 (2010)
26. Michael, J.L., Raymond, H.S., Wolfgang, M.S.: Electrospun materials for energy harvesting, conversion, and storage: a review. Pure Appl. Chem. **82**, 2137–2156 (2010)
27. Fang, J., Niu, H., Wang, H., Wang, X., Lin, T.: Enhanced mechanical energy harvesting using needleless electrospun poly(vinylidene fluoride) nanofiber webs. Energ. Environ. Sci. **6**, 2196–2202 (2013)
28. Fang, J., Shao, H., Niu, H., Lin, T.: Applications of electrospun nanofibers for electronic devices. In: Tao, X. (ed.) Handbook of Smart Textiles, pp 617–652. Springer Singapore, Singapore (2015)
29. Baji, A., Agarwal, K., Oopath, S.V.: Emerging developments in the use of electrospun fibers and membranes for protective clothing applications. Polymers **12** (2020)
30. Mwiiri, F.K., Daniels, R.: Chapter 3—Electrospun nanofibers for biomedical applications. In: Shegokar, R. (ed.) Delivery of Drugs, pp 53–74. Elsevier (2020)
31. Rasouli, R., Barhoum, A., Bechelany, M., Dufresne, A.: Nanofibers for biomedical and healthcare applications. Macromol. Biosci. **19**, 1800256 (2019)
32. Rivero, G., Aldana, A.A., Abraham, G.A.: Chapter 4—Nanocomposite electrospun micro/nanofibers for biomedical applications. In: Grumezescu, V., Grumezescu, A.M. (eds.) Materials for Biomedical Engineering, pp 89–126. Elsevier (2019)
33. Yoon, K., Hsiao, B.S., Chu, B.: Functional nanofibers for environmental applications. J. Mater. Chem. **18**, 5326–5334 (2008)
34. Weikel, A.L., Cho, S.Y., Morozowich, N.L., Nair, L.S., Laurencin, C.T., Allcock, H.R.: Hydrolysable polylactide–polyphosphazene block copolymers for biomedical applications: synthesis, characterisation, and composites with poly(lactic-co-glycolic acid). Polym. Chem. **1**, 1459–1466 (2010)
35. Swider, E., Koshkina, O., Tel, J., Cruz, L.J., de Vries, I.J.M., Srinivas, M.: Customising poly(lactic-co-glycolic acid) particles for biomedical applications. Acta Biomater. **73**, 38–51 (2018)

36. Huang, T., Zhang, Y., He, P., Wang, G., Xia, X., Ding, G., Tao, T. H.: "Self-Matched" tribo/piezoelectric nanogenerators using vapor-induced phase-separated poly(vinylidene fluoride) and recombinant spider silk. **32**, 1907336 (2020)
37. Xu, D., Zhang, H., Pu, L., Li, L.: Fabrication of Poly(vinylidene fluoride)/multiwalled carbon nanotube nanocomposite foam via supercritical fluid carbon dioxide: synergistic enhancement of piezoelectric and mechanical properties. Compos. Sci. Technol. **192**, 108108 (2020)
38. Xu, J.-F., Chen, Y.-Z., Wu, D., Wu, L.-Z., Tung, C.-H., Yang, Q.-Z.: Photoresponsive hydrogen-bonded supramolecular polymers based on a stiff stilbene unit. Angew. Chem. Int. Ed. **52**, 9738–9742 (2013)
39. Jørgensen, L., Qvortrup, K., Chronakis, I.S.: Phospholipid electrospun nanofibers: effect of solvents and co-axial processing on morphology and fiber diameter. RSC Adv. **5**, 53644–53652 (2015)
40. Nuansing, W., Georgilis, E., de Oliveira, T.V.A.G., Charalambidis, G., Eleta, A., Coutsolelos, A.G., Mitraki, A., Bittner, A.M.: Electrospinning of tetraphenylporphyrin compounds into wires. Part. Part. Syst. Charact. **31**, 88–93 (2014)
41. Ranjbar Mohammadi, M., Kargozar, S., Bahrami, S.H., Rabbani, S.: An excellent nanofibrous matrix based on gum tragacanth-poly (Ɛ-caprolactone)-poly (vinyl alcohol) for application in diabetic wound healing. Polym. Degrad. Stab. **174**, 109105 (2020)
42. Santos, L.G., Oliveira, D.C., Santos, M.S.L., Neves, L.M.G., de Gaspi, F.O.G., Mendonca, F.A.S., Esquisatto, M.A.M., Santos, G.M.T., d'Ávila, M.A., Mei, L.H.I.: Electrospun membranes of poly(Lactic Acid) (PLA) used as scaffold in drug delivery of extract of sedum dendroideum. J. Nanosci. Nanotechnol. **13**, 4694–4702 (2013)
43. Kennedy, K.M., Bhaw-Luximon, A., Jhurry, D.: Cell-matrix mechanical interaction in electrospun polymeric scaffolds for tissue engineering: implications for scaffold design and performance. Acta Biomater. **50**, 41–55 (2017)
44. Jiang, T., Carbone, E.J., Lo, K.W.H., Laurencin, C.T.: Electrospinning of polymer nanofibers for tissue regeneration. Prog. Polym. Sci. **46**, 1–24 (2015)
45. Song, J., Zhang, Y.: Vertically aligned silicon carbide nanowires/reduced graphene oxide networks for enhancing the thermal conductivity of silicone rubber composites. Compos. Part A. **133**, 105873 (2020)
46. Shah, S.A., Kulhanek, D., Sun, W., Zhao, X., Yu, S., Parviz, D., Lutkenhaus, J.L., Green, M.J.: Aramid nanofiber-reinforced three-dimensional graphene hydrogels for supercapacitor electrodes. J. Colloid Interface Sci. **560**, 581–588 (2020)
47. Ponnamma, D., Nair, S.S., Parangusan, H., Hassan, M.K., Adham, S., Karim, A., Al-Maadeed, M.A.A.: White graphene-cobalt oxide hybrid filler reinforced polystyrene nanofibers for selective oil absorption. Polymers (Basel, Switz.) **12**, 4 (2020)
48. Song, N., Pan, H., Liang, X., Cao, D., Shi, L., Ding, P.: Structural design of multilayer thermally conductive nanofibrillated cellulose hybrid film with electrically insulating and antistatic properties. J. Mater. Chem. C. **6**, 7085–7091 (2018)
49. Singh, A.K., Shishkin, A., Koppel, T., Gupta, N.: A review of porous lightweight composite materials for electromagnetic interference shielding. Compos. B **149**, 188–197 (2018)
50. Kumar, A., Prasada Rao, T., Jayakumar, O.D., Nazri, G.A., Naik, V.M., Naik, R.: Improved electrochemical performance of Li_2FeSiO_4/CNF/rGO nanocomposites for lithium ion batteries. Solid State Ionics **325**, 43–47 (2018)
51. Kato, T., Matsumoto, T., Hongo, C., Nishino, T.: Mechanical and thermal properties of cellulose nanofiber composites with nanodiamond as nanocarbon filler. Nanocomposites **4**, 127–136 (2018)
52. Verdejo, R., Bernal, M.M., Romasanta, L.J., Tapiador, F.J., Lopez-Manchado, M.A.: Reactive nanocomposite foams. Cell Polym. **30**, 45–62 (2011)
53. Akpan, E.I., Shen, X., Wetzel, B., Friedrich, K.: Design and synthesis of polymer nanocomposites. In: Pielichowski, K., Majka, T.M. (eds.) Polymer Composites with Functionalized Nanoparticles, pp 47–83. Elsevier (2019)
54. Jemima, W.S., Magesan, P., Chiranjeevi, P., Umapathy, M.J.: Sorption properties of organo modified montmorillonite clay for the reclamation of chromium (VI) from waste water. SILICON **11**, 925–933 (2019)

55. Erkliğ, A., Doğan, N.F.: Nanographene inclusion effect on the mechanical and low velocity impact response of glass/basalt reinforced epoxy hybrid nanocomposites. J. Braz. Soc. Mech. Sci. Eng. **42**, 83 (2020)
56. Chen, J., Qiu, Y., Yang, D., She, J., Wang, Z.: Improved piezoelectric performance of two-dimensional ZnO nanodisks-based flexible nanogengerators via ZnO/Spiro-MeOTAD PN junction. J. Mater. Sci. Mater. Electron. **31**, 5584–5590 (2020)
57. Ebrahimi, F., Hosseini, S.H.S.: Nonlinear dynamics and stability of viscoelastic nanoplates considering residual surface stress and surface elasticity effects: a parametric excitation analysis. Eng. Comput. (2020)
58. Hiramatsu, M., Hori, M.: Growth mechanism of carbon nanowalls. In: Hiramatsu, M., Hori, M. (eds.) Carbon Nanowalls: Synthesis and Emerging Applications, pp 81–116. Springer Vienna, Vienna (2010)
59. Gunes, S., Tamburaci, S., Tihminlioglu, F.: A novel bilayer zein/MMT nanocomposite incorporated with H. perforatum oil for wound healing. J. Mater. Sci. Mater. Med. **31**, 7 (2019)
60. Bhat, P., Ullas Pradhan, U., Naveen Kumar, S.K.: Synthesis and characterization of ZnO nano discs using wet chemical method for sensing applications. Mater. Today Proc. **5**, 10763–10770 (2018)
61. Santhosh, M.N., Filipič, G., Kovacevic, E., Jagodar, A., Berndt, J., Strunskus, T., Kondo, H., Hori, M., Tatarova, E., Cvelbar, U.: N-graphene nanowalls via plasma nitrogen incorporation and substitution: the experimental evidence. Nano-Micro Lett. **12**, 53 (2020)
62. Shah, S., Hussain, S., Qiao, G., Tan, J., Javed, M. S., Zulfiqar, Ge, C., Wang, M., Liu, G.: Decorating spherical In_2O_3 nanoparticles onto ZnO nanosheets for outstanding gas-sensing performances. J. Mater. Sci. Mater. Electron. **31**, 3924–3933 (2020)
63. Vengatesan, M.R., Mittal, V.: Nanoparticle and nanofiber-based polymer nanocomposites: an overview. In: Spherical and Fibrous Filler Composites, pp. 1–38 (2016)
64. Dresselhaus, M.S., Avouris, P.: Introduction to carbon materials research. In: Dresselhaus, M.S., Dresselhaus, G., Avouris, P. (eds.) Carbon Nanotubes: Synthesis, Structure, Properties, and Applications, pp 1–9. Springer Berlin Heidelberg. Berlin, Heidelberg (2001)
65. Kozlov, G.V., Dolbin, I.V.: Effectiveness of carbon nanotubes and graphene in reinforcement. Russ. Eng. Res. **40**, 313–315 (2020)
66. Kozlov, G.V., Dolbin, I.V.: Dependence of the degree of reinforcement of polymer/2D-nanofiller nanocomposites on the nanofiller surface structure. J. Surf. Invest. **13**, 1086–1089 (2019)
67. Kim, Y.S., Yu, B.-K., Kim, J.W., Suh, Y.-H., Kim, D.-Y., Kim, W.B.: Building a hybrid nanocomposite assembly of gold nanowires and thienyl-derivative fullerenes to enhance electron transfer in photovoltaics. J. Mater. Chem. A **1**, 5015–5020 (2013)
68. Saranya, M., Ramachandran, R., Wang, F.: Graphene-zinc oxide (G-ZnO) nanocomposite for electrochemical supercapacitor applications. J. Sci. Adv. Mater. Dev. **1**, 454–460 (2016)
69. Yang, X., Meng, N., Zhu, Y., Zhou, Y., Nie, W., Chen, P.: Greatly improved mechanical and thermal properties of chitosan by carboxyl-functionalised MoS_2 nanosheets. J. Mater. Sci. **51**, 1344–1353 (2016)
70. Costa, R.S., Moura, E.A.B.: A bibliometric analysis of the strategy and performance measurement of the polymer matrix nanomaterials development scenario globally, and the participation of Brazil. In: Li, J., Zhang, M., Li, B., Monteiro, S.N., Ikhmayies, S., Kalay, Y.E., Hwang, J.-Y., Escobedo-Diaz, J.P., Carpenter, J.S., Brown, A.D. (eds.) Characterisation of Minerals, Metals, and Materials 2020, pp. 329–342 Springer International Publishing, Cham (2020)
71. Morán, J.I., Alvarez, V.A., Cyras, V.P., Vázquez, A.: Extraction of cellulose and preparation of nanocellulose from sisal fibers. Cellulose **15**, 149–159 (2008)
72. Manikandan, D., Mangalaraja, R.V., Siddheswaran, R., Avila, R.E., Ananthakumar, S.: Fabrication of nanostructured clay–carbon nanotube hybrid nanofiller by chemical vapour deposition. Appl. Surf. Sci. **258**, 4460–4466 (2012)
73. Tomer, V., Polizos, G., Manias, E., Randall, C.A.: Epoxy-based nanocomposites for electrical energy storage. I: effects of montmorillonite and barium titanate nanofillers. J. Appl. Phys. **108**, 074116 (2010)

74. Lee, Y., Park, Y.-J., Kim, C., So, J.-H., Yeom, B., Koo, H.-J.: A conducting composite microfiber containing graphene/silver nanowires in an agarose matrix with fast humidity sensing ability. Polymer **164**, 1–7 (2019)
75. Liu, Y., Wang, Y., Zhang, C., Liu, T.: 2D nanosheet-constructed hybrid nanofillers for polymer nanocomposites with synergistic dispersion and function. APL Mater. **7**, 080904 (2019)
76. Qurashi, A., Subrahmanyam, K.S., Kumar, P.: Nanofiller graphene–ZnO hybrid nanoarchitecture: optical, electrical and optoelectronic investigation. J. Mater. Chem. C. **3**, 11959–11964 (2015)
77. Kumar, A.P., Depan, D., Singh Tomer, N., Singh, R.P.: Nanoscale particles for polymer degradation and stabilisation—trends and future perspectives. Prog. Polym. Sci. **34**, 479–515 (2009)
78. Qing, Y., Jiang, Y., Lin, H., Wang, L., Liu, A., Cao, Y., Sheng, R., Guo, Y., Fan, C., Zhang, S., Jia, D., Fan, Z.: Boosting the supercapacitor performance of activated carbon by constructing overall conductive networks using graphene quantum dots. J. Mater. Chem. A **7**, 6021–6027 (2019)
79. Asakuma, S., Okada, H., Kunitake, T.: Template synthesis of two-dimensional network of cross-linked acrylate polymer in a cast multibilayer film. J. Am. Chem. Soc. **113**, 1749–1755 (1991)
80. Evans, K.E., Alderson, A., Christian, F.R.: Auxetic two-dimensional polymer networks. An example of tailoring geometry for specific mechanical properties. J. Chem. Soc. Faraday Trans. **91**, 2671–2680 (1995)
81. McDonald, W.F., Koren, A.B.: US6951902B2 (2002)
82. Viana de Carvalho, P.H., Santos Barreto, A., Rodrigues, M.O., de Menezes Prata, V., Barreto Alves, P., de Mesquita, M.E., Alves Júnior, S., Navickiene, S.: Two-dimensional coordination polymer matrix for solid-phase extraction of pesticide residues from plant Cordia salicifolia. J. Sep. Sci. **32**, 2132–2138 (2009)
83. Li, M., Huang, X., Wu, C., Xu, H., Jiang, P., Tanaka, T.: Fabrication of two-dimensional hybrid sheets by decorating insulating PANI on reduced graphene oxide for polymer nanocomposites with low dielectric loss and high dielectric constant. J. Mater. Chem. **22**, 23477–23484 (2012)
84. Li, G., Li, Y., Liu, H., Guo, Y., Li, Y., Zhu, D.: Architecture of graphdiyne nanoscale films. Chem. Commun. **46**, 3256–3258 (2010)
85. Brunetto, G., Autreto, P.A.S., Machado, L.D., Santos, B.I., dos Santos, R.P.B., Galvão, D.S.: Nonzero gap two-dimensional carbon allotrope from porous graphene. J. Phys. Chem. C **116**, 12810–12813 (2012)
86. Schrier, J.: Fluorinated and nanoporous graphene materials as sorbents for gas separations. ACS Appl. Mater. Interfaces **3**, 4451–4458 (2011)
87. Schrier, J.: Carbon dioxide separation with a two-dimensional polymer membrane. ACS Appl. Mater. Interfaces **4**, 3745–3752 (2012)
88. Aykut, G., Hasirci, V.N., Alaeddinoglu, G.: Immobilisation of yeast cells in acrylamide gel matrix. Biomaterials **9**, 168–172 (1988)
89. Murphy, W.L., Dennis, R.G., Kileny, J.L., Mooney, D.J.: Salt fusion: an approach to improve pore interconnectivity within tissue engineering scaffolds. Tissue Eng. **8**, 43–52 (2002)
90. Sahiner, N.: In situ metal particle preparation in cross-linked poly(2-acrylamido-2-methyl-1-propanesulfonic acid) hydrogel networks. Colloid Polym. Sci. **285**, 283–292 (2006)
91. Zapadinskii, B.I., Kotova, A.V., Matveeva, I.A., Pevtsova, L.A., Stankevich, A.O., Shashkova, V.T., Barachevskii, V.A., Dunaev, A.A., Timashev, P.S., Bagratashvili, V.N.: UV-irradiation-induced formation of gold nanoparticles in a three-dimensional polymer matrix. Khim Fiz. **29**, 87–96 (2010)
92. Liu, S., Maheshwari, R., Kiick, K.L.: Polymer-based therapeutics. Macromolecules **42**, 3–13 (2009)
93. Uhrich, K.E., Abdelhamid, D.: 3—Biodegradable and bioerodible polymers for medical applications. In: Poole-Warren, L., Martens, P., Green, R. (eds.) Biosynthetic Polymers for Medical Applications, pp 63–83. Woodhead Publishing (2016)

94. Keys, K.B., Andreopoulos, F.M., Peppas, N.A.: Poly(ethylene glycol) star polymer hydrogels. Macromolecules **31**, 8149–8156 (1998)
95. Phelps, E.A., Enemchukwu, N.O., Fiore, V.F., Sy, J.C., Murthy, N., Sulchek, T.A., Barker, T.H., García, A.J.: Maleimide cross-linked bioactive PEG hydrogel exhibits improved reaction kinetics and cross-linking for cell encapsulation and in situ delivery. Adv. Mater. **24**, 64–70 (2012)
96. Wirkner, M., Alonso, J.M., Maus, V., Salierno, M., Lee, T.T., García, A.J., del Campo, A.: Triggered cell release from materials using bioadhesive photocleavable linkers. Adv. Mater. **23**, 3907–3910 (2011)
97. Wirkner, M., Weis, S., San Miguel, V., Álvarez, M., Gropeanu, R.A., Salierno, M., Sartoris, A., Unger, R.E., Kirkpatrick, C.J., del Campo, A.: Photoactivatable caged cyclic RGD peptide for triggering integrin binding and cell adhesion to surfaces. ChemBioChem **12**, 2623–2629 (2011)
98. Zheng, Y., Chen, Z., Jiang, Q., Feng, J., Wu, S., del Campo, A.: Near-infrared-light regulated angiogenesis in a 4D hydrogel. Nanoscale **12**, 13654–13661 (2020)
99. Oh, J.K., Siegwart, D.J., Lee, H.-I., Sherwood, G., Peteanu, L., Hollinger, J.O., Kataoka, K., Matyjaszewski, K.: Biodegradable nanogels prepared by atom transfer radical polymerization as potential drug delivery carriers: synthesis, biodegradation, in vitro release, and bioconjugation. J. Am. Chem. Soc. **129**, 5939–5945 (2007)
100. Li, A., Dearman, B.L., Crompton, K.E., Moore, T.G., Greenwood, J.E.: Evaluation of a novel biodegradable polymer for the generation of a dermal matrix. J. Burn. Care. Res. **30**, 717–728 (2009)
101. Kirkby, M.J.: The fractal geometry of nature. Benoit B. Mandelbrot. W. H. Freeman and co., San Francisco, 1982. Earth Surf. Proc. Land. **8**, 406–406 (1983)
102. Chandra, A.: Anion clustering and fractal pattern growth in ion conducting polymeric matrix. Solid State Ionics **86–88**, 1437–1442 (1996)
103. Chen, T.-Y., Woo, E.M., Nagarajan, S.: Periodic fractal-growth branching to nano-structured grating aggregation in phthalic acid. Sci. Rep. **10**, 4062 (2020)
104. Jäger, K.-M., McQueen, D.H.: Fractal agglomerates and electrical conductivity in carbon black polymer composites. Polymer **42**, 9575–9581 (2001)
105. Kozlov, G.V., Lipatov, Y.S.: Fractal and structural aspects of adhesion in particulate-filled polymer composites. Compos. Interfaces. **9**, 509–527 (2002)
106. Kozlov, G.V., Lipatov, Y.S.: Change in the structure of polymer matrix of particulate-filled composites: the fractal treatment. Mech. Compos. Mater. **40**, 545–550 (2004)
107. Ghanbari, K., Mousavi, M.F., Shamsipur, M., Rahmanifar, M.S., Heli, H.: Change in morphology of polyaniline/graphite composite: a fractal dimension approach. Synth. Met. **156**, 911–916 (2006)
108. Sill, T.J., von Recum, H.A.: Electrospinning: applications in drug delivery and tissue engineering. Biomaterials **29**, 1989–2006 (2008)
109. Ahmed, A., Jia, Y., Huang, Y., Khoso, N.A., Deb, H., Fan, Q., Shao, J.: Preparation of PVDF-TrFE based electrospun nanofibers decorated with PEDOT-CNT/rGO composites for piezoelectric pressure sensor. J. Mater. Sci. Mater. Electron. **30**, 14007–14021 (2019)
110. Xue, J., Wu, T., Dai, Y., Xia, Y.: Electrospinning and electrospun nanofibers: methods, materials, and applications. Chem. Rev. **119**, 5298–5415 (2019)
111. Zhu, N., Liu, W., Xue, M., Xie, Z., Zhao, D., Zhang, M., Chen, J., Cao, T.: Graphene as a conductive additive to enhance the high-rate capabilities of electrospun $Li_4Ti_5O_{12}$ for lithium-ion batteries. Electrochim. Acta **55**, 5813–5818 (2010)
112. Scaffaro, R., Lopresti, F., Maio, A., Botta, L., Rigogliuso, S., Ghersi, G.: Electrospun PCL/GO-g-PEG structures: processing-morphology-properties relationships. Compos. A Appl. Sci. Manuf. **92**, 97–107 (2017)
113. Scaffaro, R., Maio, A., Lopresti, F., Botta, L.: Nanocarbons in electrospun polymeric nanomats for tissue engineering: a review. Polymers **9** (2017)
114. Jang, W., Yun, J., Jeon, K., Byun, H.: PVdF/graphene oxide hybrid membranes via electrospinning for water treatment applications. RSC Adv. **5**, 46711–46717 (2015)

115. Kim, C.H., Kim, B.-H., Yang, K.S.: TiO_2 nanoparticles loaded on graphene/carbon composite nanofibers by electrospinning for increased photocatalysis. Carbon **50**, 2472–2481 (2012)
116. Huang, C.-L., Wu, H.-H., Jeng, Y.-C., Liang, W.-Z.: Electrospun graphene nanosheet-filled poly(trimethylene terephthalate) composite fibers: effects of the graphene nanosheet content on morphologies, electrical conductivity, crystallisation behavior, and mechanical properties. Polymers (Basel, Switz.) **11**, 164/1–164/23 (2019)
117. Zhao, J., Zhu, J., Li, Y., Wang, L., Dong, Y., Jiang, Z., Fan, C., Cao, Y., Sheng, R., Liu, A., Zhang, S., Song, H., Jia, D., Fan, Z.: Graphene quantum dot reinforced electrospun carbon nanofiber fabrics with high surface area for ultrahigh rate supercapacitors. ACS Appl. Mater. Interfaces **12**, 11669–11678 (2020)
118. Chen, Z., Jiang, Y., Xin, B., Jiang, S., Liu, Y., Lin, L.: Electrochemical analysis of conducting reduced graphene oxide/polyaniline/polyvinyl alcohol nanofibers as supercapacitor electrodes. J. Mater. Sci. Mater. Electron. **31**, 5958–5965 (2020)
119. Abdel-Mottaleb, M.M., Mohamed, A., Karim, S.A., Osman, T.A., Khattab, A.: Preparation, characterisation, and mechanical properties of polyacrylonitrile (PAN)/graphene oxide (GO) nanofibers. Mech. Adv. Mater. Struct. **27**, 346–351 (2020)
120. Asif, S.A.B., Khan, S.B., Asiri, A.M.: Efficient solar photocatalyst based on cobalt oxide/iron oxide composite nanofibers for the detoxification of organic pollutants. Nanoscale Res. Lett. **9**, 510 (2014)
121. Nho, Y.-C., Park, J.-S., Lim, Y.-M.: Preparation of poly(acrylic acid) hydrogel by radiation crosslinking and its application for mucoadhesives. Polymers **6** (2014)
122. Parani, M., Lokhande, G., Singh, A., Gaharwar, A.K.: Engineered nanomaterials for infection control and healing acute and chronic wounds. ACS Appl. Mater. Interfaces **8**, 10049–10069 (2016)
123. Mo, X.M., Xu, C.Y., Kotaki, M., Ramakrishna, S.: Electrospun P(LLA-CL) nanofiber: a biomimetic extracellular matrix for smooth muscle cell and endothelial cell proliferation. Biomaterials **25**, 1883–1890 (2004)
124. Andreu, V., Mendoza, G., Arruebo, M., Irusta, S.: Smart dressings based on nanostructured fibers containing natural origin antimicrobial, anti-inflammatory, and regenerative compounds. Materials **8** (2015)
125. Ballesteros, C.A.S., Correa, D.S., Zucolotto, V.: Polycaprolactone nanofiber mats decorated with photoresponsive nanogels and silver nanoparticles: slow release for antibacterial control. Mater. Sci. Eng. C. **107**, 110334 (2020)
126. Asadi, H., Ghaee, A., Nourmohammadi, J., Mashak, A.: Electrospun zein/graphene oxide nanosheet composite nanofibers with controlled drug release as antibacterial wound dressing. Int. J. Polym. Mater. Polym. Biomater. **69**, 173–185 (2020)
127. Alhusein, N., Blagbrough, I.S., Beeton, M.L., Bolhuis, A., De Bank, P.A.: Electrospun zein/PCL fibrous matrices release tetracycline in a controlled manner, killing staphylococcus aureus both in biofilms and ex vivo on pig skin, and are compatible with human skin cells. Pharm. Res. **33**, 237–246 (2016)
128. Mazaheri, M., Akhavan, O., Simchi, A.: Flexible bactericidal graphene oxide–chitosan layers for stem cell proliferation. Appl. Surf. Sci. **301**, 456–462 (2014)
129. Guo, X., Mei, N.: Assessment of the toxic potential of graphene family nanomaterials. J. Food Drug. Anal. **22**, 105–115 (2014)
130. Bora, C., Bharali, P., Baglari, S., Dolui, S.K., Konwar, B.K.: Strong and conductive reduced graphene oxide/polyester resin composite films with improved mechanical strength, thermal stability and its antibacterial activity. Compos. Sci. Technol. **87**, 1–7 (2013)
131. Catrina, S.-B., Okamoto, K., Pereira, T., Brismar, K., Poellinger, L.: Hyperglycemia regulates hypoxia-inducible factor-1α protein stability and function. **53**, 3226–3232 (2004)
132. Fadini, G.P., Sartore, S., Schiavon, M., Albiero, M., Baesso, I., Cabrelle, A., Agostini, C., Avogaro, A.: Diabetes impairs progenitor cell mobilisation after hindlimb ischaemia–reperfusion injury in rats. Diabetologia **49**, 3075–3084 (2006)
133. Botusan, I.R., Sunkari, V.G., Savu, O., Catrina, A.I., Grünler, J., Lindberg, S., Pereira, T., Ylä-Herttuala, S., Poellinger, L., Brismar, K., Catrina, S.-B.: Stabilisation of HIF-1α is critical to improve wound healing in diabetic mice. **105**, 19426–19431 (2008)

134. Gao, W., Sun, L., Fu, X., Lin, Z., Xie, W., Zhang, W., Zhao, F., Chen, X.: Enhanced diabetic wound healing by electrospun core–sheath fibers loaded with dimethyloxalylglycine. J. Mater. Chem. B **6**, 277–288 (2018)
135. Jo, S.B., Erdenebileg, U., Dashnyam, K., Jin, G.-Z., Cha, J.-R., El-Fiqi, A., Knowles, J.C., Patel, K.D., Lee, H.-H., Lee, J.-H., Kim, H.-W.: Nano-graphene oxide/polyurethane nanofibers: mechanically flexible and myogenic stimulating matrix for skeletal tissue engineering. J. Tissue Eng. **11**, 2041731419900424 (2020)
136. Khoramgah, M.S., Ranjbari, J., Abbaszadeh, H.-A., Tabatabaei Mirakabad, F.S., Hatami, S., Hosseinzadeh, S., Ghanbarian, H.: Freeze-dried multiscale porous nanofibrous three dimensional scaffolds for bone regenerations. Bioimpacts **10**, 73–85 (2020)
137. Yogeswari, C., Hijas, K.M., Girisun, T.C.S., Nagalakshmi, R.: Intensity dependent nonlinear absorption switching behavior of electrospun meta-nitroaniline nanofiber. Opt. Mater. **100**, 109691 (2020)
138. Kumar, P., Kiani, A.: Synthesis of electrospun nanofibrous structures with controlled optical and electrical properties. AIP Adv. **10**, 015121 (2020)
139. Li, B., Luo, J., Huang, X., Lin, L., Wang, L., Hu, M., Tang, L., Xue, H., Gao, J., Mai, Y.-W.: A highly stretchable, super-hydrophobic strain sensor based on polydopamine and graphene reinforced nanofiber composite for human motion monitoring. Compos. Part B Eng. **181**, 107580 (2020)
140. Jia, Y., Yue, X., Wang, Y., Yan, C., Zheng, G., Dai, K., Liu, C., Shen, C.: Multifunctional stretchable strain sensor based on polydopamine/reduced graphene oxide/electrospun thermoplastic polyurethane fibrous mats for human motion detection and environment monitoring. Compos. Part B Eng. **183**, 107696 (2020)
141. Mishra, R.K., Nawaz, M.H., Hayat, A., Nawaz, M.A.H., Sharma, V., Marty, J.-L.: Electrospinning of graphene-oxide onto screen printed electrodes for heavy metal biosensor. Sens. Actuators B Chem. **247**, 366–373 (2017)
142. Jiang, D., Ge, P., Wang, L., Jiang, H., Yang, M., Yuan, L., Ge, Q., Fang, W., Ju, X.: A novel electrochemical mast cell-based paper biosensor for the rapid detection of milk allergen casein. Biosens. Bioelectron. **130**, 299–306 (2019)
143. Jiang, L., Liu, Y., Liu, S., Zeng, G., Hu, X., Hu, X., Guo, Z., Tan, X., Wang, L., Wu, Z.: Adsorption of estrogen contaminants by graphene nanomaterials under natural organic matter preloading: comparison to carbon nanotube, biochar, and activated carbon. Environ. Sci. Technol. **51**, 6352–6359 (2017)
144. Feng, X., Cheng, H., Pan, Y., Zheng, H.: Development of glucose biosensors based on nanostructured graphene-conducting polyaniline composite. Biosens. Bioelectron. **70**, 411–417 (2015)

3D Graphene Nanocomposite by Electrospinning for Supercapacitor

Saptarshi Dhibar and Sudip Malik

Abstract In the present scenario, development of green and sustainable energy sources becomes an enormous challenge due to the rising environmental problems and diminution of natural energy sources like gas, coal and oil. To fulfill the requirement researchers from all over the world have typically focused on the development of various energy storage devices like accumulators, capacitors, batteries as well as supercapacitors utilizing different functional materials. Graphene, a novel type nano-sized carbon, has drawn significant research interest as an electrode material for supercapacitor due to its superior properties like high mechanical strength, better electrical conductivity, good chemical stability, excellent flexibility, as well as board electrochemical windows. For the fabrication of nanofiber, electrospinning technique has been known as the main adaptable as well as versatile technique. To enhance high energy storage capacity along with the environment impact of electrospun supercapacitors, researchers have widely established original resolution for storing and harvesting energy from nanofiber. The present chapter presents a clear overview of the electrospinning processes, the basic concept of the supercapacitor electrode materials and 3D graphene-based nanocomposites via electrospinning techniques for the electrode materials for supercapacitor.

Keywords Electrospinning · Supercapacitors · 3D graphene architectures · 3D graphene-based electrode materials

1 Introduction

The electrochemical energy storage techniques including batteries, electrochemical capacitors as well as fuel cells are opening a new impact in our lives, as nearly all the

---

S. Dhibar (✉) · S. Malik
School of Applied and Interdisciplinary Sciences, Indian Association for the Cultivation of Science, 2A & 2B Raja S. C. Mullick Road, Jadavpur, Kolkata 700032, India
e-mail: saptaaus2007@gmail.com

S. Malik
e-mail: psusm2@iacs.res.in

S. K. Tiwari et al. (eds.), *Electrospinning of Graphene*, Carbon Nanostructures,
https://doi.org/10.1007/978-3-030-75456-3_4

past sources of energies (like, wind power, hydropower, solar energy, bio-energy etc.) are intermitted in nature [1, 2]. Generally, the electrochemical capacitors also known as ultracapacitor or supercapacitors are the energy storage devices that can storage as well as release the electrical energy through very fast charge/discharge rates. These are also having higher power densities, lower maintenance cost, excellent cyclic life span and safe operation [3–6]. Due to these excellent properties, supercapacitors are potentially useful in various fields like in portable electronics, in vehicles, in memory backup systems as well as in different electronics industries.

Based on the charge storage mechanism, supercapacitor has been divided into two categories, pseudocapacitors as well as electrochemical double layer capacitors (EDLC). The pseudocapacitors materials can store the charge via reversible redox reactions taken place in between electrolytes in addition to electrode materials, involving oxidation–reduction and electrosorption reaction procedures. It is well known fact that more charge can be stored in pseudocapacitor materials as compared to EDLC materials because the electrochemical techniques can arise not only on the surface of the electrode materials but also in the bulk of the electrode materials. On the other hand, the EDLC materials can store the electric charges on the electrode surface along with the opposite charged ions on the electrolyte side via reversible ion adsorption/desorption technique. The EDLC materials exhibit exclusive power performance along with enormously long cycle lifetime because the electrostatic charge storage technique not only promises quick energy uptake and delivery, but also prevents the swelling of electrode materials. Carbonaceous materials are mostly utilized for the promising candidates for the EDLC materials due to their higher surface area, good electrochemical stability in solutions, superior conductivity etc. [7].

Nanofibers are fabricated by different synthetis techniques, such as by template synthesis of nano-structured polymers, self-assemble, phase separation as well as by electrospinning techniques [8, 9]. Out of which, more adaptable and unique methods for the fabrication of the nano and macro fibers are via electrospinning techniques that has become a universal technique in the area of nanotechnology [10, 11]. Most of the one-dimensional (1D) nanofibers with repeated tenability can be developed with the help of these electrospinning techniques. Compared to economical mechanical spinning technique for producing the microfibers, electrospinning technique largely makes use of the electrostatic repulsion among surface charges to diminish the diameter of a viscoelastic jet or a glassy filament. On the other hand, the electrospinning technique is comparatively easier as well as non-expensive which has the most significant advantages for the production of huge number of various kinds of nanofiber by this technique. In addition to this, other benefits of this technique are that it has the capability to control the diameters of the fiber, high aspect ratio, high surface-to-volume ratio in addition to pore size as non-woven fabrics. Due to these significant properties, the nanofibers fabricated by electrospinning technique have shown enormous potential in the field of electrode materials for supercapacitors. Many researches have already been performed on the fabrication of nanofiber, nanotubes, hierarchical nanomaterials along with coaxial nanocables that have used the electrode materials for the supercapacitors [12–15]. It has also noticed that day by day the number of publications has been rising on the field of nanomaterials prepared by electrospinning techniques for the supercapacitor electrode materials. This chapter is proposed the present research activities in the field of 3D graphene-based nanomaterials fabricated by the help of electrospinning techniques for the electrode materials for supercapacitor. The basic concept of supercapacitors and their classifications, fundamental of electrospinning including electrospinning process, and the nanocomposites produced via electrospinning process has elaborately discussed.

2 Supercapacitor

Supercapacitors also known as ultracapacitor or double-layer capacitor are mainly the electronic devices that have been utilized to store enormously huge amount of electrical charge in the electrical double layer formed at the interface of electrode materials and electrolytic solutions [2]. Their stored energy is typically 10 to 100 times more energy per unit volume or mass as compared to electrolytic capacitors. Due to its simpler and fast charging as well as faster delivery of charge, it has been preferred to batteries. The capacitance reached by utilizing the help of this technology can be as high as 12,000 F. The self-capacitance of the whole planet Earth is barely about 710 μF, that values is 15 million times less than as compared to the capacitance of a supercapacitor. Supercapacitors are generally polar devices, like electrolytic capacitors they have to be linked to the circuit. The charge discharge times for supercapacitors are equivalent to ordinary capacitors. Due to their comparatively low internal resistance, it is feasible to reach high charge discharge current. In case of battery, to reach fully charge state it takes several hours whereas in case of supercapacitors to reach the charge state it takes less than a few minutes. The specific power of supercapacitors is relatively high as compared to the batteries. It is observed that the specific powers of Li-ion batteries are in between 1 and 3 kW/kg and the supercapacitors have a specific power of roughly ~10 kW/kg. This high specific power helps supercapacitors in some of the important applications where quick busts of energy are required in a short period of time. Supercapacitors have excellent life cycles, generally millions of times of charge and discharge cycles. On the other hand, battery has the maximum life cycles of 500 or more than that.

As like electrolytic capacitors, the supercapacitor has been constructed and it contains two foil electrodes, a separator along with an electrolyte. Generally, the separator is placed in between the two electrodes to make sandwiched structure and finally rolls as rectangular or cylindrical in shape. The electrodes and the electrolytes which have been used for the fabrication of supercapacitors are completely dissimilar from those which have been utilized in normal electrolytic capacitors. A schematic representation of a supercapacitor has been illustrated in Fig. 1a. It stores the electrical energy inside the electrical double layer close to the surface having high surface area materials inside an electrode. At the time of charging of the supercapacitors, the anions as well as the cations are moved towards the electrode surface on the suitable surface of the cell. Differing surface charges on the electrodes equilibrium

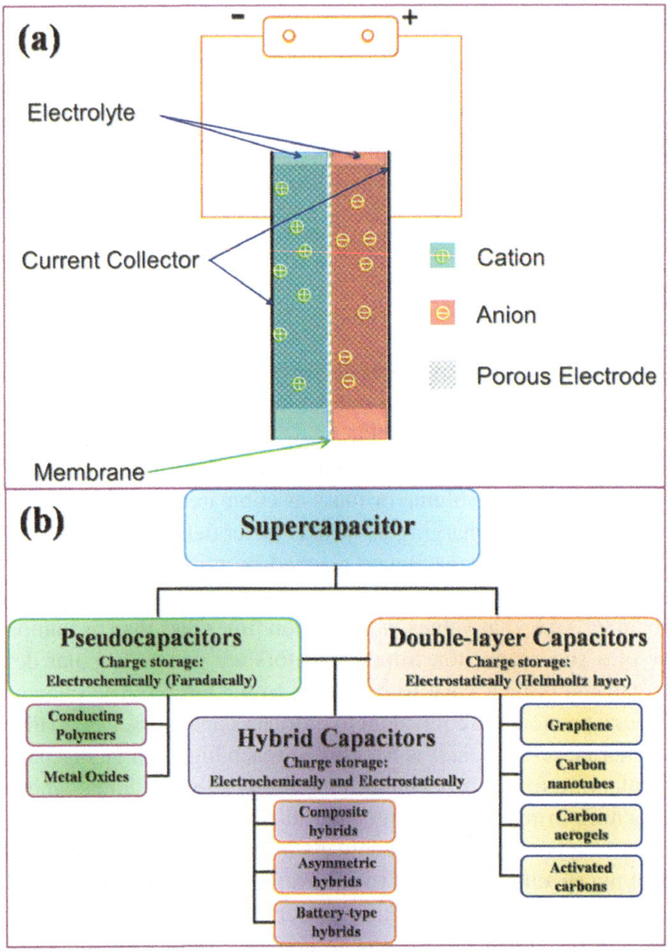

Fig. 1 **a** Schematic representation of a supercapacitor, **b** classification of supercapacitor. Reproduced from Ref. [16]

the electric field produced through the concentrated ions close to the surface, developing a potential difference among the two electrodes. At the time of discharging, the potential difference slowly reduces when ions move away from the surfaces of the electrode, and ultimately, the last result is a neutral electrolyte having least concentration gradient from the cathode and anode [16]. Concentrations of ions in the electrolyte as well as the complete active surface area inside the electrodes are the two important factors that determine the entire charge capacity along with the energy density.

Supercapacitor has numerous advantages such as, due to its low resistance it can deliver high power and allow high load current. The charging mechanism of a supercapacitor is very simple and fast and also is not cause to undergo overcharging.

As compared to batteries, a supercapacitor has marvellous low and high temperature charge along with discharge performance. Supercapacitor has also low impedance value and is extremely reliable. Apart from these good advantages, it has some disadvantages also, such as, high fabrication cost, high self-discharge, low specific energy, along with linear discharge voltage. Supercapacitors have been classified into three categories, such as pseudocapacitors, electrochemical double layer capacitors (EDLC), and hybrid capacitors. The classifications of the supercapacitors have been summarized in Fig. 1b.

2.1 Pseudocapacitors

The pseudocapacitors can store electrical energy faradaically through electron charge transfer between electrodes as well as electrolytes. This is achieved by oxidation–reduction reactions (redox reactions), electrosorption, along with intercalation processes, known as pseudocapacitance. The electrons which are engaged in the faradaic processes are transported to or from valance electron states of the redox electrode reagent. They go into the negative electrode and flow via the external circuit to the positive electrode in which a second double layer with an equivalent number of anions has produced. The electrons which arrive at the positive electrode are not moved to the anions forming the double layer, rather they can stay in the strongly ionized as well as 'electron hungry' transition metal ions of the electrode's surface.

The capability of electrodes to achieve pseudocapacitance effects by redox reactions, intercalation or electrosorption powerfully rely upon the chemical affinity of the electrode materials to the ions absorbed on the surface of the electrode along with the dimension and structure of the pores of the electrodes. Materials undergoing such redox behavior include: (i) conducting polymers, such as polyaniline, polypyrrole, polythiophene etc., (ii) transition metal oxides, like MnO_2, RuO_2, IrO_2, Co_3O_4 etc. (iii) transition metal sulfide, as well as (iv) mixed transition metal oxide and sulfides [17–26]. It is also reported that the RuO_2 is one of the most efficient electrode materials for pseudocapacitor because of its superior conductivity as well as distinct multiple reversible redox states available inside a potential window of 1.2 V. Pseudocapacitor mainly shows three types of faradaic processes such as reversible electrochemical doping-dedoping in conducting polymer based electrode, redox reactions of transition metals oxides, as well as reversible absorption (such as, adsorption of hydrogen on the surface of gold or platinum). It has also been studied that the faradaic electrochemical techniques not only enlarge the working voltages but also enhance the specific capacitance of the supercapacitors. Pseudocapacitor exhibits better specific capacitance values as well as energy density as compared to EDLC because the electrochemical processes take place both on the surface as well as in the bulk close to the surface of the solid electrode. It is also reported that the capacitance of the pseudocapacitor can be 10–100 times higher than that of electrostatic capacitance of an EDLC. Though, it suffers from comparatively lower power density compared

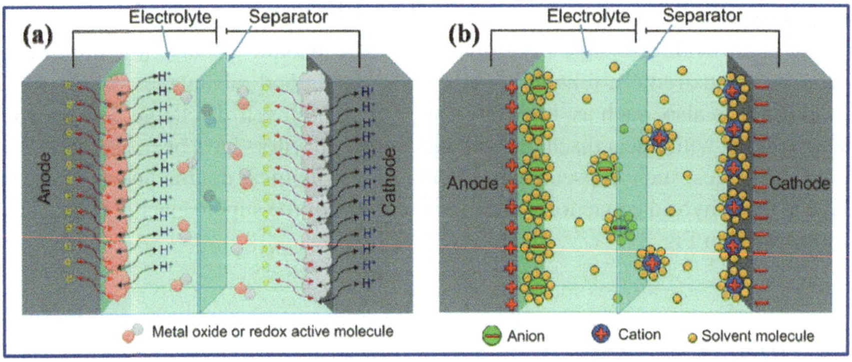

Fig. 2 Schematic representation of **a** a pseudocapacitor and **b** an electrochemical double layer capacitor. Reproduced from Ref. [27]

to EDLC as faradaic processes are generally slower than nonfaradaic processes. Further, as redox reactions take place at the electrode, a pseudocapacitor frequently lacks stability at the time of cycling. Therefore, the cyclic stability of a pseudocapacitor is less as compared to EDLC. A schematic representation of a pseudocapacitor is represented in Fig. 2a.

2.2 Electrochemical Double Layer Capacitor (EDLC)

In an electrostatic or EDLC, the capacitance of the electrode/interface is connected to an electrode-potential-dependent accumulation of charge at the interface. The mechanism of surface electrode charge generation comprises ion adsorption along with surface dissociation from both the crystal lattice defect as well as electrolytes [3]. These processes function exclusively on the electrostatic accretion of surface charge. The EDLC comes from electrode materials particles, for example at the interface among the carbon particles as well as the electrolyte, in which an excess or a shortage of electric charge is built up on the surface of the electrode, and electrolyte ions with counterbalancing charge are accumulated on the electrolyte side in order to meet the electro-neutrality. At the time of charging, the electrons are moved from the negative electrode to the positive electrode via an external load. Inside the electrolyte, anions are migrated towards the positive electrode whereas the cations are migrated towards the negative electrode. At the time of discharging, opposite processes happen. In this category of supercapacitors, there is no charge transfer in between the electrode/electrolyte, and no ion exchanges take place among the electrode as well as the electrolyte. This indicates that at the time of charging or discharging processes the electrolyte concentration remains constant. Through this procedure, the energy is stored in the EDLC [5]. Mainly carbonaceous materials such as carbon nanotubes (both multi-walled carbon nanotubes and single-walled carbon nanotubes), graphene, carbon aerogels, activated carbon, carbon nanohorns etc. show the EDLC type of behavior. However, the individual material suffers from good capacitance properties. Though, their cyclic stability is relatively very high as compared to the pseudocapacitor material.

Over the last few decades, different types of carbonaceous materials have been broadly studied as the electrode materials for EDLC [28–31]. It is reported that the activated carbon having a large BET surface area of 500–3000 m^2/g has already been utilized in commercialized EDLC [31]. For any EDLC, the electrolyte is the main key factor to determine the capacitance properties. It is also reported that the double layer capacitance of any carbonaceous materials in organic electrolytes is much lesser than that in aqueous electrolytes. The occurrence of such behavior is due to the high resistance of the organic electrolytes [3]. Though, the operating voltage in aqueous electrolyte is much lesser (hardly about 1 V) because of the limitation of the electrochemical water decomposition, while it can be enhanced in organic electrolyte to above 2.0 V and in ionic liquid electrolyte to far beyond 3.0 V [32, 33]. A schematic representation of an electrochemical double layer capacitor is represented in Fig. 2b.

2.3 Hybrid Capacitor

Another type of supercapacitor is the hybrid supercapacitor that has asymmetrical electrode configuration in which one electrode consists of pseudocapacitor (redox active) materials whereas the second electrode consists of EDLC (carbonaceous) materials. Presently, these types of hybrid supercapacitors are extensively studied to take advantage of both the electrode materials in enhancing the energy density, power density and the overall cell voltages [34–36]. In this type of hybrid supercapacitor, both the faradaic capacitance and electrochemical double layer capacitance are taking place at the same time, in which one of them plays a major role. In both types of mechanisms, huge surface area, high conductivity as well as suitable pore size distribution are the main crucial properties of the electrode materials to reach excellent specific capacitance. Apart from asymmetric electrode configuration, these types of hybrid supercapacitor also can be fabricated by the help of pseudocapacitor as well as EDLC materials in an individual composite. As we know that the EDLC materials suffer from good specific capacitance but they have the advantages of excellent cyclic stability whereas, the pseudocapacitor material suffers from good cyclic stability although they achieve excellent specific capacitance. So, the combination of two *i.e.* pseudocapacitor materials along with EDLC materials to make a hybrid composite which serves the purpose fruitfully, that means the achievement of superior specific capacitance along with excellent cyclic stability [37–40].

3 Electrospinning

Electrospinning has been recognized for almost a century, and is regarded as a variation of the electrospray process [41–44]. At the time of electrospraying techniques, the liquid drop enlarges with an enhanced electric field. While the repulsive force induced by the distribution of charge on the surface of the drop is equalized by the surface tension of the liquid, the liquid drops deforms into a conical shape. A jet of liquid ejects from the cone tip, once the repulsive force surpasses the surface tension. In the case of low viscous liquids, the small droplets form as a result of the varicose breakup of the jet. When this behavior is applied in case of polymer solutions, instead of breaking up into single drops a solid fiber is produced.

The foremost observation on electrospinning, initially known as electrospraying was studied by Rayleigh in the year 1897 and Zeleny in the year 1914 [45]. Later, in the year 1934, a patent was published by Formhals in which the first explanation of an apparatus for the electrospinning process was demonstrated. The production of textile yarn from cellulose acetate was also described in that patent [46]. Though until the early 1990s, he did not receive much attention on that. From that time, various researches were carried out in order to assemble fiber processing by the help of electrostatic forces. Afterward, in the year 1969, Taylor was studied on electrically driven jets which happen to have a strong foundation in understanding the electrospinning technique. The electrospinning term was first used in the year around 1994. Later, an enormous contribution with this electrospinning technique was made by Reneker and co-workers [47–50]. They have mainly studied the electrospinning technique for a series of different polymers in their work. The mechanism for the development of polymer fibers throughout the electrospinning technique has been demonstrated by them. In order to increase its production as well as efficiency in fabricating versatile nanofibers, the electrospinning technique has undergone many modifications and improvements nowadays. There are numerous advanced methods and setups of electrospinning machines utilized totally according to the desired characterizations.

3.1 Fundamental of Electrospinning

Electrospinning is relatively very simple, inexpensive as well as a highly adaptable way to produce polymeric nanomaterials having diameter in the range of nanometers under a high electric field [51–53]. Apart from polymeric nanomaterials, inorganic nanomaterials along with one dimensional (1D) nanocomposite can also be produced by simply adding extra ingredients into the electrospun polymeric solution or combining with post-treatment procedures. Usually, a high-voltage power supply, a spinneret (generally syringe with built-up needle) and a grounded collector are the three major components for the basic laboratory electrospinning equipment. The high-voltage power supply provides a high voltage (several tens of kilovolts) which is applied among the spinneret as well as the collector. The pendent drop of the polymer

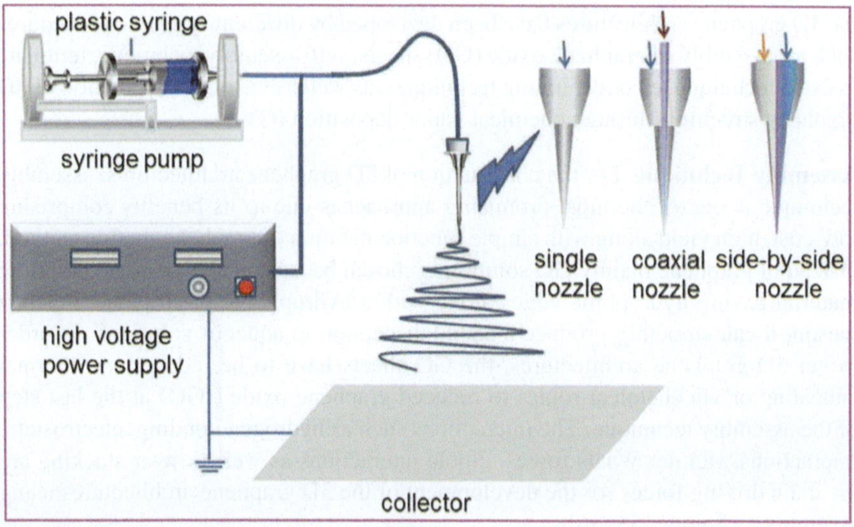

Fig. 3 Schematic diagram of the general laboratory setup for electrospinning with three different types of the nozzles. Reproduced from Ref. [54]

melt or solution is electrified and experiences the electrostatic forces under a high electric field and forms a droplet having conical shape (also known as Taylor cone) at the tip of the spinneret. When the strength of the electric field attains a critical value that means electrostatic force defeats the surface tension, a liquid jet of polymer melt or solution will eject and elongate from the tip of the Taylor cone, moving toward the electrode of the counter electrode. Then the charged jet experiences an unstable as well as whipping procedure, followed by the evaporation of the solvent or the solidification of the melt and the development of endless solid nanofibers on the counter collector. Generally, the spinneret is a syringe having a needle and utilizes as one of the electrodes of the electrospinning system. The inner diameter of a spinneret is in between 100 μm and 1 mm. A syringe pump is required to supply a controlled and constant rate to force the polymer solution to be fed through the spinneret. On the other hand, the polymer solution could be restricted in a glass tube by one end a of small diameter, otherwise a syringe without the push rod. Afterward, the syringe or the glass tube is tilted to generate the polymer solution flowing because of the gravity. A schematic diagram of the general laboratory setup for electrospinning as well as three different types of the nozzles has been highlighted in Fig. 3.

3.2 *Fabrication Techniques of 3D Graphene Architectures*

For the past few years, significant attention has been given to the development of 3D graphene materials with diverse morphologies and functionalities. It is observed that the 3D graphene architectures have been developed by different synthetic procedures such as, assembly of graphene oxide (GO) sheets, self-assembly technique, template assisted technique, electrospinning techniques, as well as via direct deposition of 3D graphene structures through chemical vapor deposition (CVD).

Assembly Technique For the construction of 3D graphene architectures, assembly technique is one of the most promising approaches due to its benefits comprising low-cost, high yield along with simple functionalization of graphene. In this method, instead of graphene mainly GO solution is chosen because it acts as an amphiphilic material having hydrophilic edges along with a hydrophobic basal plane. For this reason, it can smoothly produce a stable dispersion in aqueous solutions. In order to get 3D graphene architectures, the GO sheets have to be reduced via thermal annealing or via chemical routes to reduced graphene oxide (rGO) at the last step of the assembly technique. The interactions such as hydrogen bonding, electrostatic interactions, van der Waals forces, dipole interactions as well as π–π stacking are the main driving forces for the development of the 3D graphene architectures using assembly technique [55, 56].

Self Assembly Technique Another method for the fabrication of the 3D graphene is the self-assembly technique. In this technique mainly the 2D graphene sheets have been converted to 3D graphene architectures through different functionalities. Generally, via gelation of GO dispersion followed by reduction of GO to rGO, the 3D graphene architecture has been developed. Alternatively, by the help of hydrothermal or chemical reduction procedures the 3D graphene architectures can also be directly achieved through the self-assembly of GO sheets [57]. In these procedures, the GO sheets are easily self-assembled to form 3D architectures, and simultaneously reduced to rGO.

Template Assisted Technique Template assisted method is another suitable and feasible technique for the fabrication of 3D graphene architectures. In which pre-designed 3D templates like polystyrene, silicon dioxide has been used and after the reduction of GO it removes from the structures. Normally, the used template is bounded by graphene sheets because of the electrostatic interaction among positively charged templates as well as negatively charged graphene sheets. In this procedure more controlled morphologies and properties have been obtained. It is also observed that the size of the 3D graphene architectures directly depends on the template size. Up to now, considerable effort has already been given for the development of the 3D graphene architectures by the help of this template assisted technique [58, 59]. The 3D graphene architectures have also fabricated in another convenient method in which the assembly of GO sheets on top of the 3D templates followed by reduction of GO to rGO [57].

Electrospraying Technique The 3D graphene architectures have also been fabricated by the help of electrospinning or electrospraying techniques. Generally, the core–shell electrospinning techniques has attained great research interest because of its feasibility to achieve multi-functionality and use various materials in one step procedure simply via removing deposition steps as like self assemble as well as template assisted method and therefore it extends the prospective applications in various fields like in energy storage, nanocomposites, drug delivery, sensors etc. [60]. Poudeh et al. fabricated the 3D graphene based hollow architectures via electrospraying technique [60]. In their study, 2D GO sheets are converted into 3D hollow and filled microspheres by the help of three different polymers [polystyrene (PS), polymethyl methacrylate (PMMA), and polyacrylonitrile (PAN)] through one step core–shell electrospraying method without applying any post treatment. It has been observed that the electrospraying technique prevents the aggregations and crumbling of graphene sheets by constructing 3D interconnected framework, and gives a homogeneous dispersion of graphene sheets in polymer solution under electric field, and permits the polymer chains to crawl into graphene layers forming intercalated structure. By taking the help of Mark-Houwink-Sakurada equation, the solution viscosities as well as the proper polymer concentration are determined to fabricate an ideal graphene based polymeric sphere structure through electrospraying. They have confirmed that the PS and PMMA are utilized as carrier polymers to convert 2D graphene sheets into 3D spheres because these polymers are easily processable for bead formation and are extensively utilized as templates to construct hollow 3D structures. The schematic representation of the fabrication of graphene-based 3D spheres by tri-axial electrospraying technique has been illustrated in Fig. 4.

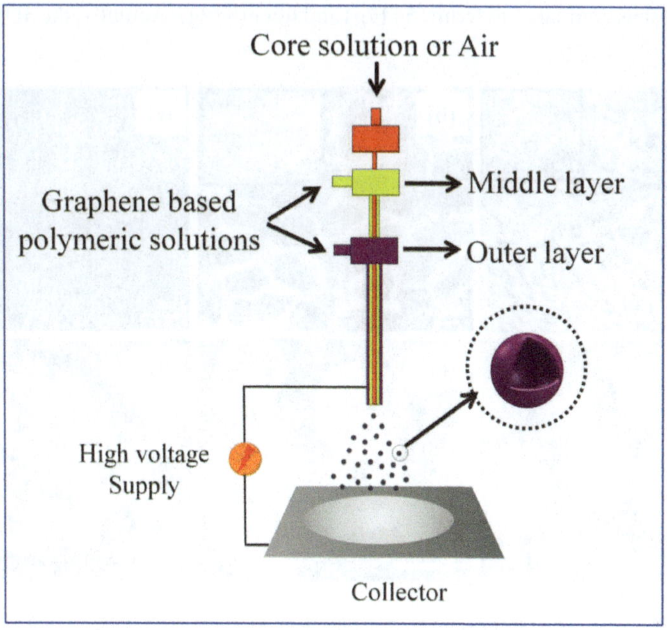

Fig. 4 Schematic illustration of the fabrication of graphene-based 3D spheres by tri-axial electrospraying technique. Reproduced from Ref. [60]

4 3D Graphene Nanocomposite for Supercapacitor

This section has been elaborately discussed about the fabrication 3D graphene-based nanocomposite by electrospinning techniques for the electrode materials for supercapacitors. Poudeh et al. successfully fabricated the platinum (Pt) – decorated 3D graphene-based carbon composite foam for the first time via core–shell electrospinning techniques instead of CVD methods [61]. In this procedure, by tailoring the concentration of polymer, applied voltage as well as polymer molecular weight, the multilayer graphene sheets are transformed into three different forms such as fiber, spheres and foam. When the polymer concentration is enhanced, it continues the generation of fibers, whereas, upon decreasing the concentration of polymer has caused the formation of graphene-based foam. Further, the reduction in molecular weight of polymer in electrospun solution has led to the construction of 3D spherical structures. All the structures 3D graphene-based spheres, foams as well as fibers have been confirmed from the morphological analysis. SEM images of electrosprayed, reduced and carbonized Pt-decorated 3D graphene-based foams as well as spheres are illustrated in Fig. 5. The electrochemical characteristic of the Pt-decorated 3D graphene-based structures has been investigated by cyclic voltammogram (CV) with the help of three-electrode method using 0.5 M H_2SO_4 as electrolyte, using Ag/AgCl as reference and platinum wire as reference electrode. The specific capacitance value was calculated from the cyclic voltammogram (CV) analysis. The Pt-decorated 3D graphene-based spheres exhibit the specific capacitance of 118 F/g at 1 mV/s scan rate, highest as compared to foam (54 F/g) and fiber (8 F/g). Actually, the 3D graphene

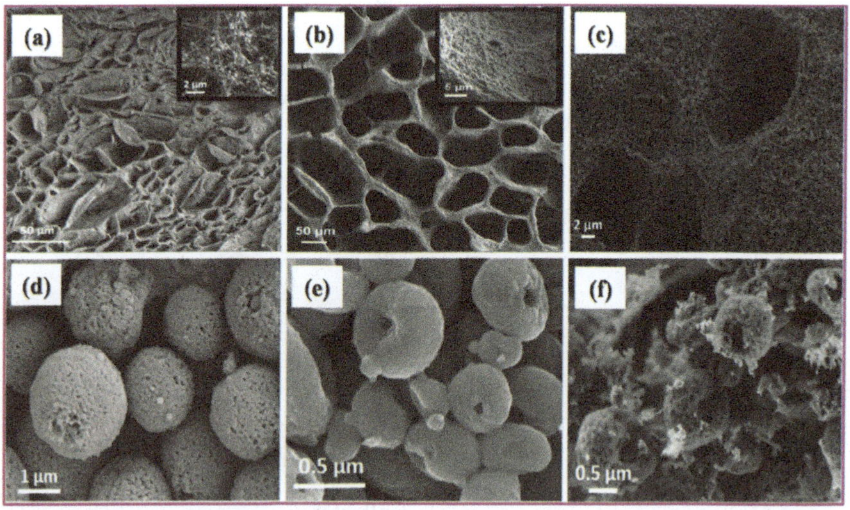

Fig. 5 SEM images of **a** electrosprayed, **b** reduced, and **c** carbonized Pt-decorated 3D graphene foam; SEM images of **d** electrosprayed, **e** reduced, and **f** carbonized Pt-decorated 3D graphene spheres. Reproduced from Ref. [61]

structures consist of micro and macro interconnected pores, which produces high surface area along with fast ion/electron transport channels. As the diameter of the Pt-nanoparticles is considerably high (~50 nm), hence the electrochemical behavior of Pt-decorated 3D graphene-based foam decreases, even if it has a unique structure. In the case of 3D graphene-based sphere, Pt-nanoparticles size decreased to 5 nm and the structure became more porous, thus improving the electrochemical properties. Their CV curves as well as GCD profile have been depicted in Fig. 6. It was observed that at low scan rate the CV curves are rectangular in shape indicating EDLC behavior, whereas the rectangular shape starts to distort on enhancing the scan rate up to 100 mV/s. Further, it has been observed from GCD profile that the Pt-decorated 3D graphene-based spheres reveal much higher discharge time compared to other structures indicating the better specific capacitance. Later the cyclic stability has been studied for all the three structures and the 3D graphene-based spheres retain their 88% of specific capacitance after 1000 charge–discharge cycles, while the foam and fibers maintain their 94 and 90% capacitance after the same charge–discharge cycles. The small diminish in cyclic stability of spheres stems from a rapid shrinking of pores of spheres because of their 3D structure, which effects in a decrease in the capacitance as time passes.

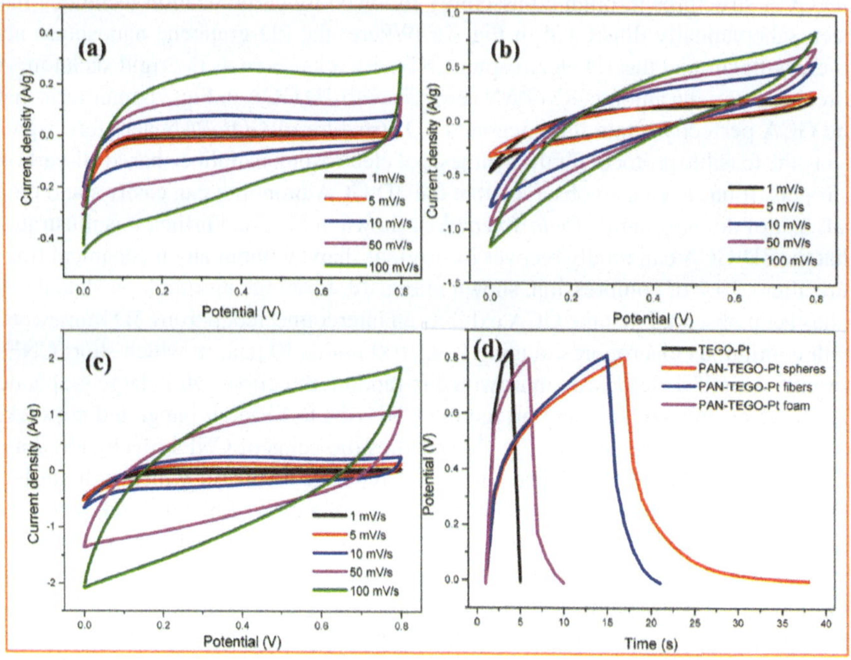

Fig. 6 CV curves at different scan rates for **a** fibers, **b** foam, **c** spheres, along with **d** GCD profiles at a 2 A/g current density for the Pt-decorated 3D graphene-based structures. Reproduced from Ref. [61]

The same groups have also fabricated the 3D urchin-shaped hybrid composite electrodes with PANI coated graphene and manganese oxide (Mn_3O_4) embedded in carbon fiber (CF) via core shell electrospinning, heat treatment followed by *in-situ* polymerization technique [62]. The electrospinning process was performed with an applied voltage of 12 kV, flow rate of 9 μL/min and having the nozzle to collector distance of 10 cm while atmospheric air was utilized as core material. Morphological analysis confirms the formation of 3D urchin-shaped PANI/Mn_3O_4/Graphene nanoplatelet (GNP) composite. The electrochemical performances were carried out by the help of a three-electrode system using 1 M H_2SO_4 as aqueous electrolyte. It has been observed that the 3D urchin-shaped PANI coated Mn_3O_4/GNP-CF composite produces the highest specific capacitance of 452 F/g at 1 mV/s scan rate and also retain their 89% of specific capacitance after 1000 charge–discharge cycles. The introduction of GNP into the fiber structures helped both ions along with electron transport and gave highly electroactive sites that shortened the diffusion paths by bridging the adjacent individual PANI chains.

Further, Huang et al. reported an efficient surface-induced co-assembly strategy for the novel design and reconstruction of electrospun nanofibers into 3D graphene/carbon nanofiber (CNF) composite aerogels [GCA] having hierarchical structures for the efficient energy storage material [63]. The preparation pathway of 3D GCA structures through co-assembly followed by carbonization technique has been schematically illustrated in Fig. 7a. Where, the 2D graphene nanosheets act as cross-linker and the 1D electrospun CNF networks serve as the rigid skeleton. A direct comparison among GO/oPAN aerogels with 3D GCA in Fig. 7b manifests that the GCA perfectly retain the integrated 3D structure of GO/oPAN aerogels, signifying the feasible protocol for the synthesis of electrospun nanofiber-based all-carbon aerogels. It has been also observed that the 3D GCA monolith can easily stand on a soft hair of dog's tail grass (*Setaria viridis*), shown in Fig. 7c. Further, it is illustrated that the 3D GCA can totally recover its original shape without any mechanical fracture after ~70% of compression, shown in Fig. 7d. From the morphological analysis it has been observed that the GCA exhibits an interconnected, porous 3D framework with continuous macropores ranging from 100 nm to 10 μm, in which short CNFs serving as rigid skeletons are intertwined in random directions while large graphene flakes act as cross-linkers and bridge the CNF ribs to form an integrated network, shown in Fig. 7e. The inset of Fig. 7e specifies the veins of CNFs ribs by red dots, revealing that CNFs are bridged to form connected pathways which will further improve the structural stability of GCA. The combination of hierarchical 3D structures having numerous macropores, with high ion-accessible surface area of 57.8 m^2/g and interconnected electrically conductive network creates GCA extremely desirable for electrochemical behaviors. They have performed the electrochemical characteristics by a three-electrode system using 6 M KOH aqueous electrolyte. The CV curves of the 3D GCA structures show nearly rectangular in shape signifying a typical EDLC behavior and fast charging/discharging processes, shown in Fig. 8a. Figure 8b represents the GCD plots at various current densities of the 3D GCA structures and conformed longer galvanostatic charge–discharge time as compared to CNFs. It has been observed that the 3D GCA structures exhibited the highest

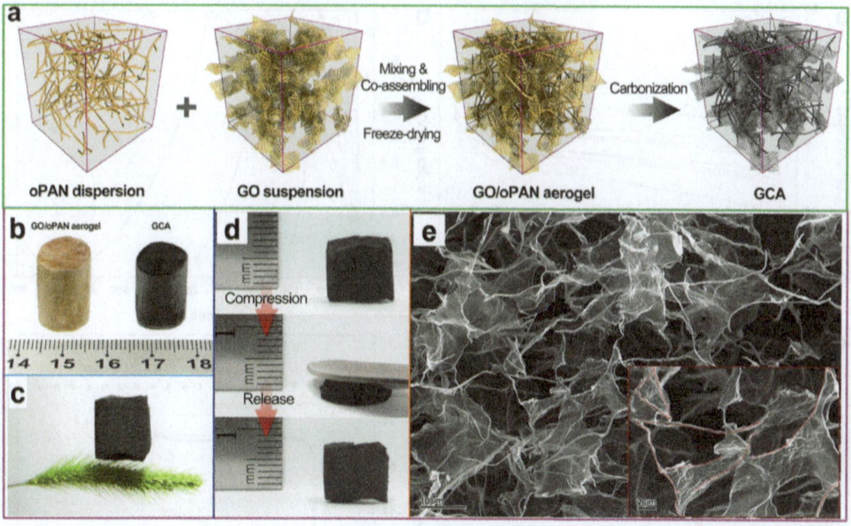

Fig. 7 **a** Schematic representation of the preparation pathway of the 3D GCA structure by co-assembly and carbonization; demonstration on the structural stability and elasticity of GCA with low density, and characterization on the interconnections among graphene and CNF with GCA: **b** GCA maintains the good cylindrical shape of GO-oPAN aerogels without noticeable shrinkage, **c** a GCA cylinder standing on the soft hair of dog's tail grass (*Setaria viridis*), and **d** GCA aerogel can recover its original shape after a strong compression. **e** SEM image of the microscopically porous architecture of GCA, with the inset showing the CNF framework by red dots. Reproduced from Ref. [63]

specific capacitance of 180 F/g at 1 A/g current density, much higher as compared to CNFs (76 F/g at A/g). The specific capacitance values at different current densities are summarized in Fig. 8c. Later, the cyclic stability analysis was carried out and it was observed that the 3D GCA structures retained their 94.8% of specific capacitance after 2000 charge–discharge cycles, shown in Fig. 8d.

An activated carbon nanowhisker (ACNWs) wrapped-on graphitized electrospun nanofiber (GENF) network [ACNWs/GENFN] having 3D porous structures was prepared by the help of electrospinning techniques followed by in-situ polymerization as a new type of binder-free electrode materials for supercapacitors was studied by He et al. [64]. The digital images of the as prepared GENF mat (gray in color) along with PANI/GENF mat (green in color) are shown in Fig. 9a. From the morphological analysis it was observed that a kind of highly porous 3D network material composited by a large amount of GENFs with smooth surface having diameter of 450–600 nm, illustrated in Fig. 9b and the inset of Fig. 9b displayed an ordered graphitic structure, which implied a high electrical conductivity. The green color mat was achieved after the growth of a layer of ordered needle-like PANI nanowhiskers (PANI-NWs) on the GENFs. Figure 9c exhibited the SEM images of 3D PANI-NWs/GENF composite fibers and observed a coaxial structure having an outer diameter of about 1.2 μm and length of about 300 nm. When the PANI-NWs are being treated at high temperature

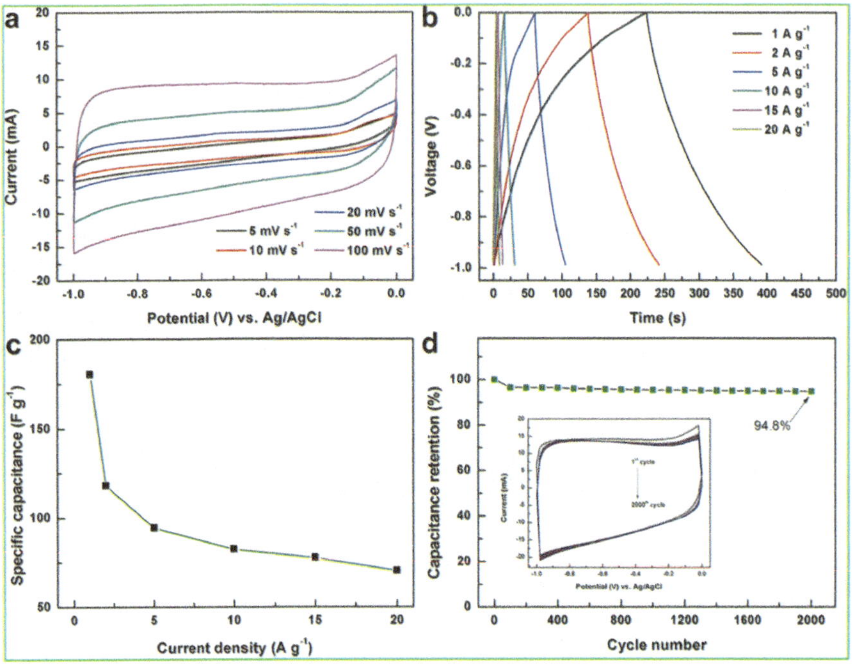

Fig. 8 Capacitive tents for 3D GCA electrodes: **a** CV curves at different scan rates, **b** GCD curve at various current densities, **c** specific capacitance at different current densities, and **d** cyclic stability test up to 2000 charge–discharge cycles, with inset showing the CV curve of the 1st cycle and CV curves of the 2000 cycles. Reproduced from Ref. [63]

it is converted to carbon nanowhiskers (CNWs), shown in Fig. 9d. Further increase in the carbonization temperature, the CNWs on the surface of GENFs become sparser and blunter, illustrated in Fig. 9e. The electrochemical performances were carried out in a three-electrode measurement system using 6 M KOH solution as electrolyte. The 3D coaxial ACNWs/GENF network composite exhibited the highest specific capacitance of 176.5 F/g at a current density of 800 A/g, ultrahigh power density of 250 kW/kg and good cyclic stability without any capacitance decreases even after 10,000 charge/discharge cycles at 100 A/g. Such a good rate capacitance performance of the ACNWs/GENFN was attributed to the 3D porous architecture, high electrically conductive network of GENFN and high utilization rate of porous ACNWs.

In another study, 3D interconnected networks of reduced graphene oxide (rGO)/thorn-like TiO_2 nanofiber (TTF) aerogels [GTTF aerogels] having different TTF weight ratios were successfully prepared by electrospinning, silica etching followed by hydrothermal combination technique [65]. From the morphological analysis of the GTTF-20 aerogels it was confirmed that the interconnected porous 3D structure has a pore diameter of several micrometers. Later, all the electrochemical tests were performed by the help of a three-electrode cell using 1 M Na_2SO_4 as the electrolyte. The working electrode has been fabricated by directly placing a small

Fig. 9 **a** Digital images of GENF mat (gray) and PANI/GENFs mat (green), **b** SEM images of GENFs, inset TEM image of GENFs, **c** PANI-NWs/GENFs, **d** C700 [CNWs/GENFs carbonized at 700 °C], **e** AC-700-600 [sample C700 activated at 600 °C]. Reproduced from Ref. [64]

piece of aerogels (~10 mm in diameter and ~1.5 mm in height) among two pieces of nickel foam (with 3 cm^2 in area) without any additives followed by pressing at 10 MPa to make the electrode materials adhere to the current collector more completely. From the GCD curve it was examined that the curves are linear and symmetrical in nature over the entire potential range, depicted in Fig. 10a. Their triangular shape also indicates excellent capacitive behavior. The GCD curve of the 3D interconnected structure based GTTF-20 aerogels at different current densities (1–5 A/g) has been illustrated in Fig. 10b. Figure 10c showed the obtained specific capacitance values at various current densities for the GTTF based 3D structures. The 3D interconnected structure based GTTF-20 aerogels exhibited the highest specific capacitance of 178 F/g at a 1 A/g current density and also maintain their 90.3% of specific capacitance after 3000 charge discharge cycles at a current density of 5 A/g, Fig. 10d. Inset of Fig. 10d showed GCD curve of the 1st as well as the 3000th cycle.

So, up to this it was seen that the development of 3D graphene-based architectures by the help of electrospinning techniques followed by other techniques tremendously enhances the ion-accessible surface area and thereby significantly improves the electrochemical characteristics. It was also observed that some of the graphene-based nanocomposites fabricated via electrospinning techniques considered as the superior electrode materials for supercapacitors. Like, Tai et al. fabricated a flexible and freestanding carbon nanofiber/graphene nanosheet (CNF/GNS) composite paper by the help of electrospinning technique followed by high-temperature annealing using

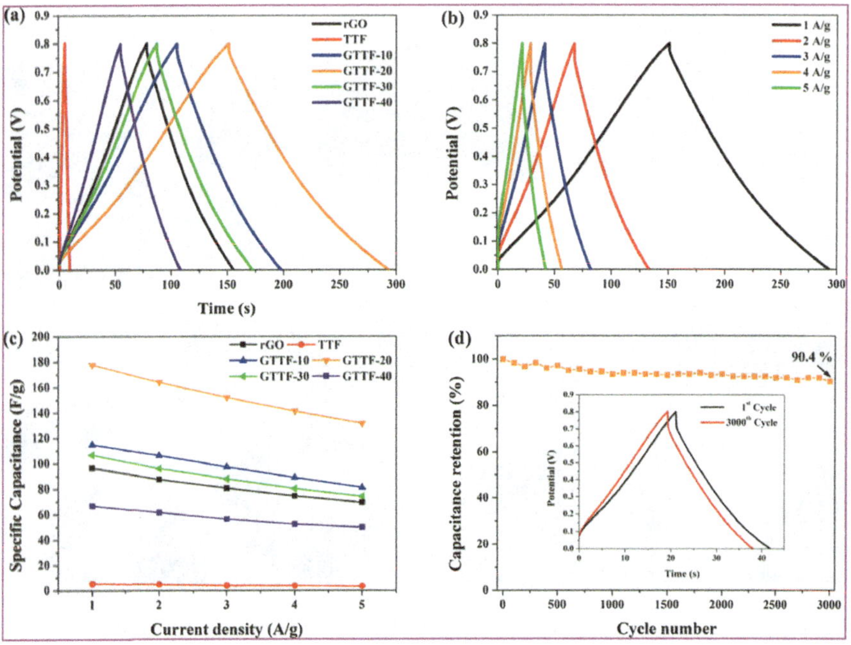

Fig. 10 **a** GCD curves of 3D GTTF aerogel-based electrodes at a current density of 1 A/g, **b** GCD curves of the GTTF-20 electrode at different current densities, **c** comparative specific capacitances of rGO and 3D GTTF aerogels, and **d** cycle performance of the 3D GTTF-20 electrode. Reproduced from Ref. [65]

a polyacrylonitrile/GNS/dimethylformamide mixture as electrospun precursor [66]. The as fabricated CNF/GNS paper showed excellent specific surface area of 447 m^2/g and an average pore size of 8.16 nm along with superior electrical conductivity of 200 S/cm. The electrochemical properties of the paper like freestanding CNT/GNS composites were carried out by the help of a three-electrode system in 6 M KOH electrolyte at room temperature. The as fabricated freestanding CNF/GNS paper achieved the highest specific capacitance of 197 F/g at a 1.25 A/g current density which is about 24% higher as compared to the pure CNF paper. The composite paper also retains its 84% of specific capacitance after 1500 charge–discharge cycles.

In another study, electrospun carbon nanofiber/graphene (CNF/G) composite was prepared by the help of *in-situ* electrospinning polymeric nanofibers with simultaneous spraying of graphene oxide, followed by heat treatment [67]. It was observed that the freestanding CNF web acts as a framework for sustaining graphene, which helps to prevent the agglomeration of graphene and to provide a high conductivity for the efficient charge transfer to the pores. The electrochemical measurements were carried out in a three-electrode electrochemical cell using 6 M KOH as electrolyte at 25 °C. The CNF/G composite achieved the highest specific capacitance of 183 F/g at a 1 A/g current density which is approximately 1.6 times high as compared to the pristine CNF. The greater specific capacitance can be attributed not only to large channels among graphene and CNF, which functions as reservoir for more electrolyte throughout the charging-discharging process, but also to large pores in the CNF/G composite, which accelerate the ionic transportation in pores. Later, the cyclic stability tests were performed at 1 A/g current density and it was observed that the CNF/G composite maintained their 92% of specific capacitance after 4500 charge–discharge cycles. Further, a nanocomposite having the combination of graphene nanoplatelets and carbon nanofibers were successfully fabricated by using a one-step solution based on the electrospinning technique for the potential electrode materials for supercapacitor [68]. The electrochemical performances were evaluated using two-electrode configuration in which liquid electrolyte socked nylon membrane filter was sandwiched between the prepared samples before being tightly fitted into an electrochemical cell. The as mentioned nanocomposite exhibited the specific capacitance of 86.11 F/g as well as retained their 90% of capacitance even after 1000 charge–discharge cycles.

In another work, the graphene oxide/polyaniline/polyvinylidene fluoride (GO/PANI/PVDF) composite nanofiber was fabricated by the help of electrospinning techniques as the electrode materials for supercapacitors reported by Rose et al. [69]. The electrochemical characteristics were performed at room temperature in a three-electrode set up using 1 M H_2SO_4 as electrolyte. The specific capacitance value was calculated by the help of CV curve and it was observed that the GO based composite nanofiber achieved the highest specific capacitance of 170.63 F/g, energy density of 13.92 W/kg, and density of 316.36 W/kg, respectively. The same research group also developed the GO/PANI/PVA composite nanofiber via electrospinning process for the supercapacitor electrode materials [70]. Here, all the parameters for the electrochemical characterization remain the same. The composite nanofiber reached the specific capacitance of 438.8 F/g. It was also reported that the development of flexible all-solid-state supercapacitor based on free standing, binder-free carbon nanofiber@polypyrrole@graphene (CNTs@PPy@rGO) core-double-shell film fabricated by carbonization of electrospun PAN nanofiber network followed by electrochemical deposition of PPy and then rGO layer coating [71]. The electrochemical measurements for the all-solid-state flexible supercapacitor were carried out by two-electrode setup using PVA/H_3PO_4 gel electrolyte. It was observed that the flexible all-solid-state supercapacitor device showed the highest specific capacitance of 188 F/g at the scan rate of a 2 mV/s and also retained their 59.5% of specific capacitance after 10,000 consecutive charge–discharge cycles. In order to better understand the underlying mechanism behind the excellent electrochemical characteristics of the all-solid-state flexible supercapacitor, the electrochemical performance of the CNFs@PPy@rGO electrode was further evaluated in three-electrode setup using 3 M KCl solution as aqueous electrolyte. In this case the core-double-shell electrode achieved the specific capacitance of 336.2 F/g at the scan rate of a 2 mV/s and also holds 98% of capacitance after 2500 charge–discharge cycles. Further, the GO reinforced electrospun CNF was developed for the ultrathin supercapacitor electrode [72]. Upon the addition of functionalized GO into a polymeric solution and subjected to an electrospinning technique, non-woven random nanofiber embedded with GO sheets was obtained. Their electrochemical properties were evaluated via a two-electrode system using 6 M KOH as electrolyte. The mentioned composite electrode showed the specific capacitance of 140.10 F/g at a 1 A/g current density and also retained their 96.2% of specific capacitance after vigorous 1000 charge–discharge cycles. In another study, graphene-beaded CNF (G/CNF) electrode was prepared by electrospinning PAN/N,N-dimethylformamide (DMF) solution dispersed with oxidized graphene nanosheets, followed by carbonization at 800 °C inside the tubular quartz furnace as the superior electrode materials for supercapacitor [73]. In this case, the electrochemical characterization was performed in a two-electrode system utilizing 6 M KOH as electrolyte. The G/CNF electrode researched the specific capacitance of 263.7 F/g at a 100 mA/g current density and also retained its 86.9% of specific capacitance after 2000 charge–discharge cycles.

Qiu et al. synthesized nitrogen-doped active carbon fiber/reduced graphene oxide (ACF-rGO-N) nanocomposite by electrospinning, calcinations, self-assemble reduction in addition to post treatment nitrogen doping procedure [74]. From the morphological analysis it was observed that the rGO nanosheets are not only dispersed homogeneously on the surface of the ACF, but also embedded into the ACF framework, forming the line-plane structure with the non-covalent bonding. It was also examined that the N-doping post treatment has no significant effect on the morphology of ACF-rGO. Initially, the electrochemical performance was measured in a three-electrode system in which both acidic (1 M H_2SO_4) and basic (6 M KOH) electrolytes were utilized. Later, it was studied in a quasi-solid state two-electrode system in which PVA/KOH gel electrolyte has been used which also acted as a separator for the solid-state device. It was reported that the in 6 M KOH electrolyte the ACF-rGO-N nanocomposite exhibited the highest specific capacitance of 283 F/g at 0.2 A/g current density, whereas in 1 M H_2SO_4 the nanocomposite achieved the specific capacitance of 301 F/g at the same current density. The cyclic stability of the ACF-rGO-N nanocomposite was further studied in two different electrolytes and observed that capacitance retention of 98% in 6 M KOH and 96% in 1 M H_2SO_4 electrolyte after consecutive 1000 charge–discharge cycles. Afterword, the solid-state symmetric supercapacitor device was characterized and it was observed from CV curves that there were no significant redox peaks at various scan rates indicating the typical double-layer capacitive behavior, shown in Fig. 11a. From the GCD curves it was noticed the isosceles triangles at different current densities with no distinct voltage drop, illustrated in Fig. 11b. The solid-tate symmetric device achieved the high specific capacitance of 254 F/g at a 0.2 A/g current density. The specific capacitance values at different current densities are represented in Fig. 11c. The symmetrical supercapacitor also exhibited the energy density of 35.2 Wh/kg as well as power density of 399.1 W/kg at a 0.2 A/g current density. Lastly, the symmetric supercapacitor also achieved excellent cyclic stability with 99% capacitance retention after 10,000 charge–discharge cycles, shown in Fig. 11d.

Similarly, by the help of electrospinning technique the GO/vanadium pentoxide (V_2O_5) nanofiber was prepared for the supercapacitor electrode materials [75]. Lastly, it was annealed at 550 °C to get the final composite. The electrochemical behaviors were carried out in a three-electrode cell setup using two aqueous electrolytes of

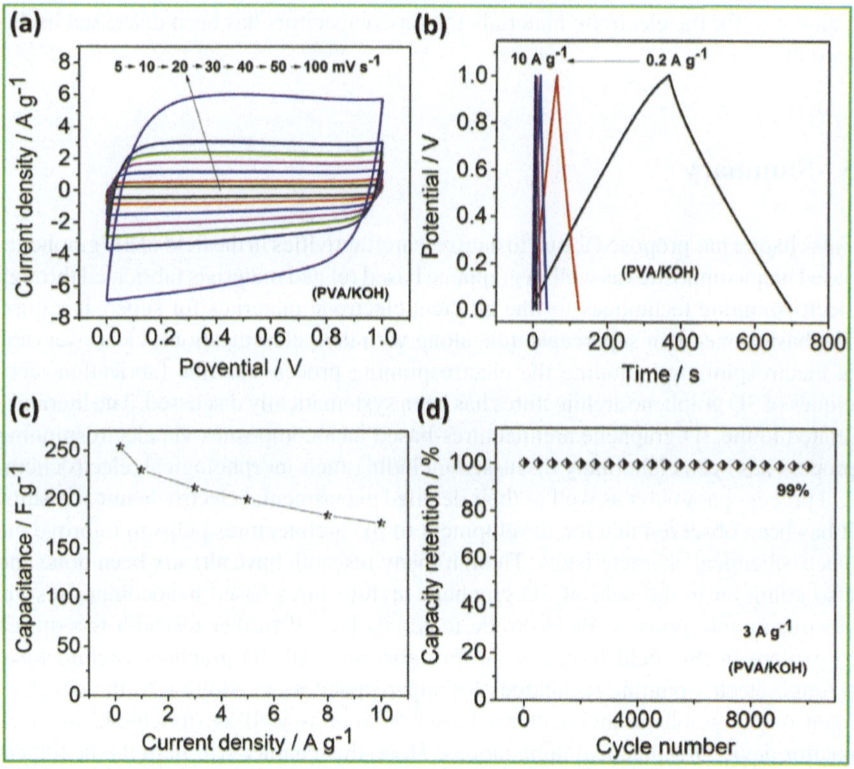

Fig. 11 Electrochemical performances of the ACF-rGO-N based solid state symmetric supercapacitor: **a** CV curved at various scan rates, **b** GCD curves different current densities, **c** specific capacitance values at various current densities, **d** cycling stability plots at current densities of 3 A/g. Reproduced from Ref. [74]

2 M KOH and 1 M H_2SO_4. The annealed GO based electrospun nanofiber exhibited the highest specific capacitance of 453.824 F/g at the scan rate of 10 mV/s in 2 M KOH electrolyte. In another study, composite of MnO_2 decorated on hierarchical porous carbon nanofiber/graphene (MnO_2/HPCNF/G) have been fruitfully fabricated via one-step electrospinning technique followed by thermal treatment as superior electrode materials for electrochemical supercapacitor [76]. The electrochemical behaviors were studied in a two-electrode test using 6 M KOH electrolyte in which the two symmetric MnO_2/HPCNF/G composite electrode were directly utilized as electrode without any polymer binder. The MnO_2/HPCNF/G composite electrode exhibited the highest specific capacitance of 210 F/g at 1 mA/cm^2 current density. It has been also observed that the MnO_2/HPCNF/G composite electrode achieved the excellent energy density of 24 Wh/kg, high power density of 400 W/kg along with superior cyclic stability as it retained 95.7% of specific capacitance after 1000 charge–discharge cycles. So, comparatively all the literature that has highlighted the 3D graphene-based nanocomposites fabricated by the help of electrospinning

techniques for the electrode materials for supercapacitors has been discussed in this section.

5 Summary

This chapter has proposed the up to date research activities in the field of 3D graphene-based nanocomposites as well as graphene based related materials fabricated through electrospinning techniques for the efficient electrode materials for supercapacitors. The basic concept of supercapacitors along with their classifications, clear overview of electrospinning including the electrospinning process and the fabrication techniques of 3D graphene architectures has been systematically discussed. The literature related to the 3D graphene architectures-based nanocomposites via electrospinning process has been elaborately discussed including their morphological, electrochemical process parameter as well as their detailed experimental electrochemical results. It has been observed that the development of 3D architectures helps to improve the electrochemical characteristics. Though many research have already been done and also going on in the field of 3D graphene architectures based nanocomposites via electrospinning process for electrode materials but till further research is required to explore in this field basically on the fabrication of 3D graphene architectures through electrospinning techniques for supercapacitors in addition to the development of 3D graphene architectures based flexible as well as stretchable supercapacitor device for practical applications. Hope this chapter will help the beginners who are working in the particular field of 3D graphene-based nanocomposites by electrospinning techniques for the electrode materials for supercapacitors.

Acknowledgements S. Dhibar would like to thank Council of Scientific & Industrial Research (CSIR), Human Resource Development Group, New Delhi, India, for the CSIR Research Associate Fellowship [File No. 09/080(1105)/2019-EMR-I]. Dr. Malik acknowledges CSIR, INDIA (Project No. 01(2875)/17/EMR-II) for the financial support.

References

1. Liu, C., Li, F., Ma, L.P., Cheng, H.M.: Advanced materials for energy storage. Adv. Mater. **22**, E28–E62 (2010)
2. Winter, M., Brodd, R.J.: What are batteries, fuel cells and supercapacitors? Chem. Rev. **104**, 4245–4270 (2004)
3. Simon, P., Gogotsi, Y.: Materials for electrochemical capacitors. Nat. Mater.**7**, 845–854 (2008)
4. Service, R.F.: New 'supercapacitor' promises to pack more electrical punch. Science **313**, 902 (2006)
5. Wang, G., Zhang, L., Zhang, J.: A review of electrode materials for electrochemical supercapacitors. Chem. Soc. Rev. **41**, 797–828 (2012)
6. Miller, J.R., Simon, P.: Electrochemical capacitors for energy management. Science **321**, 651–652 (2008)

7. Gogotsi, Y., Presser, V.: Carbon Nanomaterials. CRC Press, Boca Raton, FL (2006)
8. Kumar, A., Deka, M.: Nanofiber reinforced composite polymer electrolyte membranes. InTech (2010)
9. Ramakrishna, S., Fujihara, K., Teo, W.-E., Lim, T.-C., Ma, Z.: An Introduction to Electrospinning and Nanofibers. World Scientific Publishing Co Pte Ltd., National University of Singapore (2005)
10. Jeong, J.S., Moon, J.S., Jeon, S.Y., Park, J.H., Alegaonkar, P.S., Yoo, J.B.: Mechanical properties of electrospun PVA/MWCNTs composite nanofibers. Thin Solid Films **515**, 5136–5141 (2007)
11. Almecija, D., Blond, D., Sader, J.E., Coleman, J.N., Boland, J.J.: Mechanical properties of individual electrospun polymer-nanotube composite nanofiber. Carbon **47**, 2253–2258 (2009)
12. Chen, S., He, S., Hou, H.: Electrospinning technology for applications in supercapacitors. Curr. Org. Chem. **17**, 1402–1410 (2013)
13. Cavaliere, S., Subianto, S., Savvch, I., Jones, D.J., Rozière, J.: Electrospinning: designed architectures for energy conversion and storage devices. Energ. Environ. Sci. **4**, 4761–4785 (2011)
14. Thavasi, V., Singh, G., Ramakrishna, S.: Electrospun nanofiber in energy and environmental applications. Energ. Environ. Sci. **1**, 205–221 (2008)
15. Dong, Z.X., Kennedy, S.J., Wu, Y.Q.: Electrospinning materials for energy-related applications and device. J. Power Sour. **196**, 4886–4904 (2011)
16. Qi, Z., Koenig, G.M.: Flow battery systems with solid electroactive materials. J. Vac. Sci. Technol. B **35**, 040801–040827 (2017)
17. Conway, B.E.: Electrochemical Supercapacitors. Kluwer Academic/Plenum Press, New York (1999)
18. Wu, M.S., Chiang, P.C.J.: Fabrication of nanostructured manganese oxide electrodes for electrochemical capacitors. Electrochem. Solid-State Lett. **7**, A123–A126 (2004)
19. Sugimoto, W., Iwata, H., Murakami, Y., Takasu, Y.: Electrochemical capacitor behavior of layered ruthenic acid hydrate. J. Electrochem. Soc. **151**, A1181–A1187 (2004)
20. Wu, M.-S., Lee, R.-H., Jow, J.-J., Yang, W.-D., Hsieh, C.-Y., Weng, B.-J.: Nanostructured iron oxide films prepared by electrochemical method for electrochemical capacitors. Electrochem. Solid-State Lett. **12**, A1–A4 (2009)
21. Meher, S.K., Rao, G.R.: Ultralayered Co_3O_4 for high performance supercapacitor applications. J. Phys. Chem. C **115**, 15646–15654 (2011)
22. Xia, X.-H., Tu, J.-P., Wang, X.-L., Gu, C.-D., Zhao, X.-B.: Mesoporous Co_3O_4 monolayer hollow-sphere array as electrochemical pseudocapacitor material. Chem. Commun. **47**, 5786–5788 (2011)
23. Shen, L., Yu, L., Wu, H.B., Yu, X.-Y., Zhang, X., Lou, X.W. (David).: Formation of nickel cobalt sulfide ball-in-ball hollow spheres with enhanced electrochemical pseudocapacitive properties. Nat. Commun. **6**, 6694 (2015)
24. Xiao, J., Wan, L., Yang, S., Xiao, F., Wang, S.: Design hierarchical electrodes with highly conductive $NiCo_2S_4$ nanotube arrays grown on carbon fiber paper for high-performance pseudocapacitors. Nano Lett. **14**, 831–838 (2014)
25. Huang, L., Chen, D., Ding, Y., Feng, S., Wang, Z.L., Liu, M.: Nickel–cobalt hydroxide nanosheets coated on $NiCo_2O_4$ nanowires grown on carbon fiber paper for high-performance pseudocapacitors. Nano Lett. **13**, 3135–3139 (2013)
26. Tomboc, G.M., Jadhav, H.S., Kim, H.: PVP assisted morphology-controlled synthesis of hierarchical mesoporous $ZnCo_2O_4$ nanoparticles for high-performance pseudocapacitor. Chem. Eng. J. **308**, 202–213 (2017)
27. Chen, X., Paul, R., Dai, L.: Carbon-based supercapacitors for efficient energy storage. Natl. Sci. Rev. **4**, 453–489 (2017)
28. Béguin, F., Presser, V., Balducci, A., Frackowiak, E.: Carbons and electrolytes for advanced supercapacitors. Adv. Mater. **26**, 2219–2251 (2014)
29. Frackowiak, E.: Carbon materials for supercapacitor applications. Phys. Chem. Chem. Phys. **9**, 1774–1185 (2007)

30. Pandolfo, A.G., Hollenkamp, A.F.: Carbon properties and their role in supercapacitors. J. Power Sour. **157**, 11–17 (2006)
31. Zhang, L.L., Zhao, X.S.: Carbon-based materials as supercapacitor electrodes. Chem. Soc. Rev. **38**, 2520–2531 (2009)
32. Fernandez, J.A., Arulepp, M., Leis, J., Stoeckli, F., Centeno, T.A.: EDLC performance of carbide-derived carbons in aprotic and acidic electrolytes. Electrochim. Acta **53**, 7111–7116 (2008)
33. Chen, R., Wu, F., Li, L., Xu, B., Qiu, X., Chen, S.: Novel binary room-temperature complex system based on LiTFSI and 2-oxazolidinone and its characterization as electrolyte. J. Phys. Chem. C **111**, 5184–5194 (2007)
34. Ma, S.-B., Nam, K.-W., Yoon, W.-S., Yang, X.-Q., Ahn, K.-Y., Oh, K.-H., Kim, K.-B.: A novel concept of hybrid capacitor based on manganese oxide materials. Electrochem. Commun. **9**, 2807–2811 (2007)
35. Zhou, Y., Lachman, N., Ghaffari, M., Xu, H., Bhattacharya, D., Fattahi, P., Abidian, M.R., Wu, S., Gleason, K.K., Wardle, B.L., Zhang, Q.M.: A high performance hybrid asymmetric supercapacitor *via* nano-scale morphology control of graphene, conducting polymer, and carbon nanotube electrodes. J. Mater. Chem. A **2**, 9964–9969 (2014)
36. Gao, Z., Song, N., Li, X.: Microstructural design of hybrid CoO@NiO and graphene nano-architectures for flexible high performance supercapacitors. J. Mater. Chem. A **3**, 14833–14844 (2015)
37. Sahoo, S., Dhibar, S., Hatui, G., Bhattacharya, P., Das, C.K.: Graphene/polypyrrole nanofiber nanocomposite as electrode material for electrochemical supercapacitor. Polymer **54**, 1033–1042 (2013)
38. Dhibar, S., Das, C.K.: Silver nanoparticles decorated polyaniline/multi-walled carbon nanotubes nanocomposite for high performance supercapacitor electrode. Ind. Eng. Chem. Res. **53**, 3495–3508 (2014)
39. Dhibar, S., Bhattacharya, P., Hatui, G., Sahoo, S., Das, C.K.: Transition metal doped polyaniline/single walled carbon nanotubes nanocomposite an efficient electrode material for high performance supercapacitor. ACS Sustain. Chem. Eng. **2**, 1114–1127 (2014)
40. Dhibar, S.: Electrochemical behaviour of graphene and carbon nanotubes based hybrid polymer composites. In: Thakur, V.K., Thakur, M.K., Gupta, R.K. (eds.) Hybrid Polymer Composite Materials: Processing, pp. 211–248. Woodland Publishing, United Kingdom (2017)
41. Rayleigh, L.: On the equilibrium of liquid conducting masses charged with electricity. Philos. Mag. **14**, 184–186 (1882)
42. Cooley, J.F.: Apparatus for electrically dispersing fluids. US patent 692,631 (1902)
43. Morton, W.J.: Methods of dispersing fluids. US patent 705,691 (1902)
44. Wilson, C.T.R., Taylor, G.I.: The bursting of soap-bubbles in a uniform electric filed. Proc. Camb. Phil. Soc. **22**, 728–730 (1925)
45. Bhardwaj, N., Kundu, S.C.: Electrospinning: a fascinating fiber fabrication technique. Biotechnol. Adv. **28**, 325–347 (2010)
46. Formhals, A.: Process and apparatus for preparing artificial threads. US Patent 1975504 (1934)
47. Doshi, J., Srinivasan, G., Reneker, D.: A novel electrospinning process. Polym. News **20**, 206–207 (1995)
48. Fang, X., Reneker, D.H.: DNA fibers by electrospinning. J. Macromol. Sci. Phys. B **36**, 169–173 (1997)
49. Fong, H., Reneker, D.H.: Elastomeric nanofibers by styrene-butadiene-styrene triblock copolymer. J. Polym. Sci. Part B Polym. Phys. **37**, 3488–3493 (1999)
50. Yarin, A.L., Koombhongse, S., Reneker, D.H.: Bending instability in electrospinning nanofibers. J. Appl. Phys. **89**, 3018–3026 (2001)
51. Greiner, A., Wendorff, J.H.: Electrospinning: a fascinating method for the preparation of ultrathin fibers. Angew. Chem., Int. Ed. **46**, 5670–5703 (2007)
52. Li, D., Xia, Y.: Electrospinning of nanofibers: reinventing the wheel? Adv. Mater. **16**, 1151–1170 (2004)

53. Lu, X., Wang, C., Wei, Y.: One-dimensional composite nanomaterials: synthesis by electrospinning and their applications. Small **5**, 2349–2370 (2009)
54. Lu, X., Wang, C., Favier, F., Pinna, N.: Electrospun nanomaterials for supercapacitor electrode: designed and architectures and electrochemical performance. Adv. Energ. Mater. **7**, 1601301 (2017)
55. Bai, H., Li, C., Wang, X., Shi, G.: On the gelation of graphene oxide. J. Phys. Chem. C **115**, 5545–5551 (2011)
56. Xu, Y., Sheng, K., Li, C., Shi, G.: Self-assembled graphene hydrogel via a one-step hydrothermal process. ACS Nano **4**, 4324–4330 (2010)
57. Cao, X., Yin, Z., Zhang, H.: Three-dimensional graphene materials: preparation, structures and application in supercapacitors. Energ. Environ. Sci. **7**, 1850–1865 (2014)
58. Zhang, J., Yu, Y., Liu, L., Wu, Y.: Graphene-hollow PPy sphere 3D-nanoarchitecture with enhanced electrochemical performance. Nanoscale **5**, 3052–3057 (2013)
59. Cai, D., Ding, L., Wang, S., Li, Z., Zhu, M., Wang, H.: Facile synthesis of ultrathin-shell graphene hollow spheres for high-performance lithium-ion batteries. Electrochim. Acta **139**, 96–103 (2014)
60. Poudeh, L.H., Okan, B.S., Zanjani, J.S.M., Yildiz, M., Menceloglu, Y.Z.: Design and fabrication of hollow and filled graphene-based polymeric spheres via core-shell electrospraying. RSC Adv. **5**, 91147–91157 (2015)
61. Poudeh, L.H., Letofsky-Papst, I., Cebeci, F.Ç., Menceloglu, Y., Yildiz, M., Okan, B.S.: Facile synthesis of single- and multi-layer graphene/Mn_3O_4 integrated 3D urchin-shaped hybrid composite electrodes by core-shell electrospinning. ChemNanoMat **5**, 792–801 (2019)
62. Poudeh, L.H., Cakiroglu, D., Cebeci, F.C., Yildiz, M., Menceloglu, Y.Z., Okan, B.S.: Design of Pt-supported 1D and 3D multilayer graphene-based structural composite electrodes with controlled morphology by core-shell electrospinning/electrospraying. ACS Omega **3**, 6400–6410 (2018)
63. Huang, Y., Lai, F., Zhang, L., Lu, H., Miao, Y.-E., Liu, T.: Elastic carbon aerogels reconstructed from electrospun nanofibers and graphene as three-dimensional network matrix for efficient energy storage/conversion. Sci. Rep. **6**, 31541 (2016)
64. He, S., Chen, L., Xie, C., Hu, H., Chen, S., Hanif, M., Hou, H.: Supercapacitor based on 3D network of activated carbon nanowhiskers wrapped-on graphitized electrospun nanofiber. J. Power Sour. **243**, 880–886 (2013)
65. Kim, T.-W., Park, S.-J.: Synthesis of reduced graphene oxide/thron-like titanium dioxide nanofiber aerogels with enhanced electrochemical performances for supercapacitor. J. Colloid Interface Sci. **486**, 287–295 (2017)
66. Tai, Z., Yan, X., Lang, J., Xue, Q.: Enhancement of capacitance performance of flexible carbon nanofiber paper by adding graphene nanosheets. J. Power Sour. **199**, 373–378 (2012)
67. Dong, Q., Wang, G., Hu, H., Yang, J., Qian, B., Ling, Z., Qiu, J.: Ultrasound-assisted preparation of electrospun carbon nanofiber/graphene composite electrode for supercapacitors. J. Power Sour. **243**, 350–353 (2013)
68. Chee, W.K., Lim, H.N., Zainal, Z., Harrison, I., Andou, Y., Huang, N.M., Altarawneh, M., Jiang, Z.T.: Electrospun graphene nanoplatelets-reinforced carbon nanofibers as potential supercapacitor electrode. Mater. Lett. **199**, 200–203 (2017)
69. Rose, A., Raghavan, N., Thangavel, S., Maheswari, B.U., Nair, D.P., Venugopal, G.: Investigation of cyclic voltammetry of graphene oxide/polyaniline/polyvinylidene fluoride nanofibers prepared via electrospinning. Mater. Sci. Semicon. Process. **31**, 281–286 (2015)
70. Rose, A., Prasad, K.G., Sakthivel, T., Gunasekaran, V., Maiyalagan, T., Vijayakumar, T.: Electrochemical analysis of graphene oxide/polyaniline/polyvinyl alcohol composite nanofibers for supercapacitor applications. Appl. Surf. Sci. **449**, 551–557 (2018)
71. Chen, L., Chen, L., Ai, Q., Li, D., Si, P., Feng, J., Zhang, L., Li, Y., Lou, J., Ci, L.: Flexible all-solid-state supercapacitors based on freestanding, binder-free carbon nanofibers@polypyrrole@graphene film. Chem. Eng. J. **334**, 184–190 (2018)
72. Chee, W.K., Lim, H.N., Andou, Y., Zainal, Z., Hamra, A.A.B., Harrison, I., Altarawneh, M., Jiang, Z.T., Huang, N.M.: Functionalized graphene oxide-reinforced electrospun carbon nanofibers as ultrathin supercapacitor electrode. J. Energ. Chem. **26**, 790–798 (2017)

73. Zhou, Z., Wu, X.-F.: Graphene-beaded carbon nanofibers for use in supercapacitor electrodes: synthesis and electrochemical characterization. J Power Sour. **222**, 410–416 (2013)
74. Qiu, X., Xiao, Z., Wang, L., Fan, L.-Z.: High rate integrated quasi-solid state supercapacitors based on nitrogen-enriched active carbon fiber/reduced graphene oxide nanocomposite. Carbon **130**, 196–205 (2018)
75. Thangappan, R., Kalaiselvam, S., Elayaperumal, A., Jayavel, R.: Synthesis of graphene oxide/vanadium pentoxide composite nanofibers by electrospinning for supercapacitor applications. Solid State Ion. **268**, 321–325 (2014)
76. Lee, D.G., Kim, B.-H.: MnO_2 decorated on electrospun carbon nanofiber/grapheme composites as supercapacitor electrode materials. Synth. Met. **219**, 115–123 (2016)

Polymer Graphene-Based Nanofibers and Their Application for Batteries

Sasmita Mishra, Sandip K. Singh, and Ajeet Singh

Abstract Polymer graphene-based nanofibers are increasingly used in energy storage devices, especially in lithium-ion batteries and thus attracting more and more industrial and academic research in material chemistry. Metal-doped graphene-based polymer nanofibers are now gaining much attention and recently are successfully used in related energy storage devices, including supercapacitors and more. This chapter primarily focuses on a brief of nanofibers, electrospinning and graphene followed by the review of polymer graphene-based nanofibers in the present scenario.

Keywords Polymer graphene-based nanofibers · Lithium-ion battery · Electrospinning · Electrochemical processes · Energy storage devices

1 Introduction

Global energy needs jumped manifolds due to the rapid growth of the population. And renewable sources like solar and wind energies to meet the demand are not utilized to the fullest. Also, electrochemical energy devices like batteries are another power source that can be used and always needed to improve their level so that they can work at their highest level. High energy/power densities of Lithium-ion batteries (LIBs), Lithium-air batteries (LABs), Lithium-sulfur batteries (LSBs), Sodium-ion batteries (NIBs) and the supercapacitors (SCs) are the devices of great interest in today's scenario [1, 2]. Carbon substrate like graphene is significant due to its excellent physicochemical properties like high surface area, chemical stability, electrical conductivities and more. These excellent physicochemical properties make graphene an active electrode material [3, 4]. It can also be used as the conducting additive [5, 6] and equipment for supporting metal hydroxides and/or oxides [7–9]. Energy can

S. Mishra · A. Singh (✉)
Discipline of Chemistry, Indian Institute of Technology Indore, Simrol, Khandwa Road, Indore 453552, India

S. K. Singh
Catalysis and Inorganic Chemistry Division, CSIR- National Chemical Laboratory, Dr. Homi Bhabha Road, Pune 411 008, India

S. K. Tiwari et al. (eds.), *Electrospinning of Graphene*, Carbon Nanostructures,
https://doi.org/10.1007/978-3-030-75456-3_5

also be stored by using different allotropes of carbon, having different dimensions or properties such as high electrical conductivity along with a more significant number of active sites for physical and chemical interactions, high surface area and more. The carbon material can bear vigorous agitation without getting disintegrated; it can also provide better transportation properties as compared to several adsorbates. The best allotrope is graphene.

Graphene is a two dimensional (2D) lightweight hexagonal sheet, honeycomb-like material [10, 11]. It is considered as the superior material because of its high mechanical strength, conductivity, capacitance and other thermal, physical, chemical and biological properties [10]. It covers various fields of application which includes photonics, optoelectronics, spintronics, flexible electronics, biomedical applications and for the use focussed here, i.e. energy storage and energy generation [12–14].

Most of the research so far is carried out on the lithium-ion batteries, which consist of graphite anode and $LiCoO_2$ cathode and a polymer membrane immerse in the electrolyte. Graphite anode is a low-cost material and possesses a long cyclic life. But, using graphite based LIBs have a significant concern that is a safety issue in vehicles because of deposition of lithium on the anode surface, it also possesses a low theoretical capacity of 372 $mAhg^{-1}$[15]. In contrast, other carbons materials like carbon nanotubes (CNTs), carbon nanofibers (CNFs) and graphene show higher capacities, along with various metal oxides or metals like SnO_2, Si, Fe_2O_3 [16].

2 Fabrication of Nanofibers

The fabrication of NFs can be achieved by two main techniques, chemical vapour deposition (CVD), and spinning of carbon precursors [17, 18]. The NFs grown by CVD is called vapour grown carbon fibres (VGCFs). These were first grown using carbon filaments on metal crucibles using hydrocarbon gas in 1889 [19]. At higher temperatures ranging 700–1200 K in the presence of a metal catalyst of iron, nickel or cobalt, the diffusion of carbon atoms from hydrocarbon gas such as methane, ethane, carbon monoxide forms the fibre structure with an incorporated graphene layer [20]. The property of excellent conductivity of nanofibers because of higher graphite content, they were used as a conductive additive in place of carbon black on LIB, the electrochemical performance was found to be improved using them, but the cost of their production was high. Moreover, due to their hydrophobic surface, these were not uniformly attached, so they were not suitable for the energy storage devices [21]. (Please note, paragraph is changed here).

In comparison to graphite, carbon nanofibres were found to be superior for lithium-ion batteries application. For example, Neat $Li_4Ti_5O_{12}$(LTO) showed a lower electrical conductivity and the Li-ion diffusion coefficient as compared to LTO-CNF-1 and LTO-CNF-2 (Fig. 1). Urchin morphology of LTO-CNF-1 was more suitable due to the formation of an extensive 3D network as compared to the corn dog morphology of LTO-CNF-1. As a result of which electrical conductivity and Li-ion diffusion coefficient of the LTO-CNF-1 was much higher as compared to LTO-CNF-2.

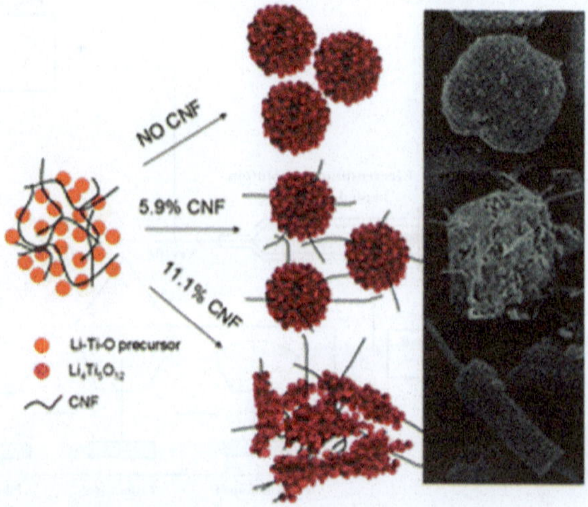

Fig. 1 Schematic diagram of the morphology of neat$Li_4Ti_5O_{12}$ (LTO), LTO–CNF-1 having 5.9% CNF and LTO–CNF-2 with 11.1 CNF composites ([21] Reproduced from Ref. with permission from The Royal Society of Chemistry)

Spinning is another technique for the preparation of nanofibers; it is of different types such as wet, gel, melt and dry spinning. In wet spinning, a polymer is solubilized in a liquid medium and, this solution is then poured in a coagulation bath through spinning, which results in the precipitation to form the solid one-dimensional (1D) polymer fibres. If the same solution is sprinkled in air and is solidified by the gas flow, it is called dry spinning [22]. In gel spinning, the solution is in gel form [23], and if the solution is melted using temperature up to 400 °C, then the technique is known as a melt spinning [24, 25]. Polyacrylonitrile (PAN) and mesophase pitch (MP) are the two necessary starting materials to produce nanofibers by spinning wet or melt. The strength of PAN-based nanofibers is comparatively high, whereas the modulus of MP based fibres is high [26]. Such nanofibres generally have a diameter of 5 μm or more. But for the production of NFs, the diameter should be less than 1 μm. Large diameter nanofibers have the disadvantage of lesser diffusion of electrons and ions into the central core part if used in energy storage devices. Its use in conductive networks also needs higher content for proper percolation [27]. In recent times, the spinning process is dominated by electrospinning.

3 Electrospinning Technique

In electrospinning, the interaction of charged polymer and external electric field controls the ejection after applying a higher voltage to the polymeric solution. The benefit of electrospinning is that the NFs formed in this case, range from ten nanometers to hundreds of nanometers, which can be utilized in various electrochemical storage systems [28]. Electrospinning setup consists of three parts: A high voltage source, spinneret and a collector or rotating drum (Fig. 2) [29]. In this technique, the

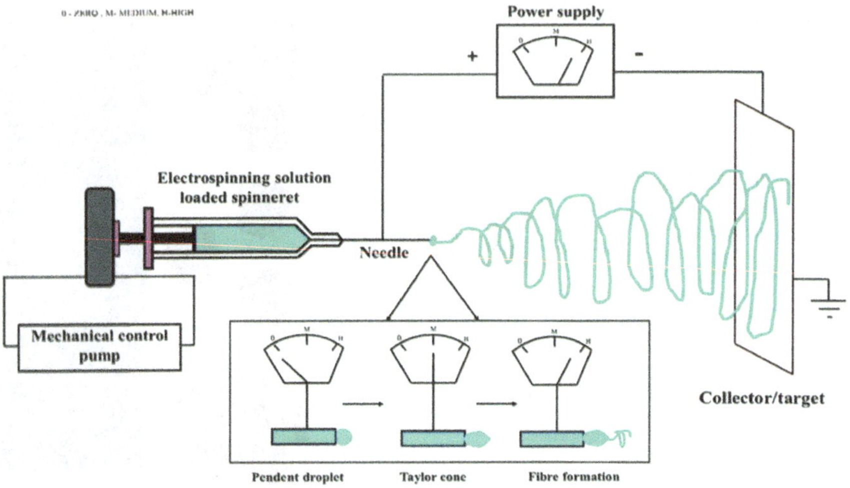

Fig. 2 Electrospinning setup showing a high voltage source, spinneret and a collector (Reproduced from Ref. [30] with permission from The Royal Society of Chemistry)

electric field is applied on the drop which hangs on the tip of the spinneret. As much as the electric field intensity increases the drop elongates and forms the cone-shaped well known as a "Taylor cone". At a particularly critical point, the surface tension of the drop is overcome by the applied field, a jet of solution accelerates downfield and electrohydrodynamic whipping instability between collector and tip stretches the liquid and its evaporation on the collector surface gives nanofibers. Its operational ease, low cost and high yield with greater reproducibility make it a priority over chemical vapour deposition.

Modification of the engineering of collectors and spinnerets can prepare different kinds of fibres with random, aligned, twisted and core/shell-like structures. The fibres with random orientation in the NFs film are mostly used in the electrochemical storage devices. The other significant kind is core/shell NFs which have metal or metal oxide as core and carbon shell. It can bear large strains generated during electrochemical reactions inside the energy devices. For example, silicon and tin NPs were used as the core with a carbon shell as anodes in LIBs [31, 32].

4 Types of Electrospinning Techniques

Various active agents, like metal nanoparticles, drugs, can be incorporated or arranged outside the nanofibers using different electrospinning techniques. The techniques include blend electrospinning, coaxial electrospinning and emulsion electrospinning (Fig. 3).

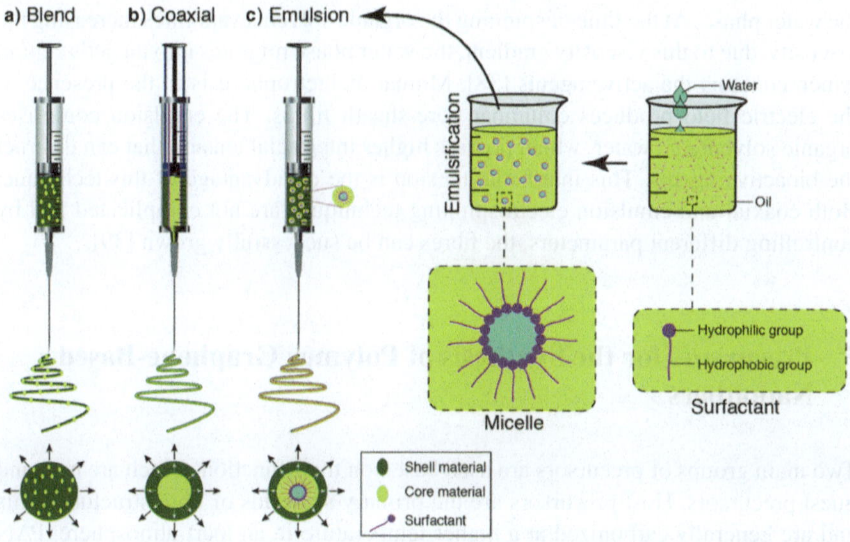

Fig. 3 Schematic representation of the spinneret loaded with the bioactive agent for **a** blend, **b** coaxial and **c** emulsion electrospinning (Reproduced from Ref. [33] with permission from The Royal Society of Chemistry)

In the blend electrospinning technique, as the name suggests, agents are blended that is dissolved or dispersed in the solution. This technique is more feasible than coaxial or emulsion electrospinning techniques. The distribution of the agents can be affected by the physicochemical properties of the solution due to their interaction with the solution [34]. Agents like proteins can be denatured during this process. In the context of the distribution of the bioactive molecules, it is proposed since the biomolecules are charged, the repulsion between charges will distribute them on the surface of the jet, due to which surface enriched fibres will be obtained instead of uniformly distributed fibres [35].

Co-electrospinning or coaxial electrospinning is the technique in which two nozzles are set up with a high voltage source that produces core–shell nanofiber. The sensitive agents are usually kept inside the inner jet and another solution is injected through the outer jet, these solutions remain unmixed till the end. Notably, both components solvents are the same to obtain the best quality nanofiber. Encapsulation of the guest molecules such as cellular scaffolds, drug molecules can be achieved successfully using this technique. Nanofluidic structures and various multiple channel nanotubes can be obtained easily [36]. The nanofibers obtained from these techniques are of core-sheath morphology. Disadvantages related to this technique involves the complexity of the design; there should be a precise control over viscoelasticity and interfacial tension to obtain the desired nanofibers [37].

Emulsion electrospinning technique uses two immiscible solutions, mostly organic solvents and water. In this technique, one component is dissolved in organic solvents (continuous phase), whereas active ingredients are dissolved in water to form the water phase. At the time of spinning the organic solvent evaporates increasing the viscosity, due to this viscosity gradient, the water phase migrates into the jet's centre, which contains the active agents [38]. Mutual dielectrophoresis in the presence of the electric field produces columnar core-sheath fibres. The emulsion comprises organic solvent and water, which possess higher interfacial tension that can destruct the bioactive agents. This interfacial tension is the disadvantage of this technique. Both coaxial and emulsion electrospinning techniques are not complicated and by controlling different parameters, the fibres can be successfully grown [39].

5 Precursors for the Synthesis of Polymer Graphene-Based Nanofibers

Two main groups of precursors are used based on their function, which are host and guest precursors. Host precursors are the primary scaffolds of fibre structural units and are generally carbonized at a higher temperature in an inert atmosphere. PAN and polyimide (PI) are two primary host precursors [40, 41]. Guest precursors as the name suggests are guests and can be removed after the heat treatment in the final NFs. In this way, voids space can be created in NFs to accommodate cores of metal/ metal oxide. PMMA and polystyrene (PS) are two main guest precursors [42, 43].

Graphene-based electro conductive electrospun nanofibers are nowadays getting a lot of attention due to their properties of being lightweight, durable, possessing high conductivity and stability. Polyacrylonitrile (PAN), and poly(methyl methacrylate) (PMMA), polyvinyl alcohol (PVA), polyvinyl chloride (PVC), nylon, poly(lactic-co-glycolic acid) (PLGA), and poly(vinyl acetate) (PVAc) are the various types of precursors that are used for the preparation of GBEENs. Matsumoto et al. prepared graphene nanoribbon (GNR)/carbon composite by using PAN containing graphene oxide nanoribbons (GONRs) followed by twisting and carbonization. They reported that the addition of GONRs improved mechanical strength as well as the electrical conductivity of 165 S cm^{-1} which was higher than PAN-based pristine CNF yarns (77 S cm^{-1}) and the monolayer GNRs (54 S cm^{-1}) [44]. Massoumi*et al.* utilized graphene oxide (GO) with poly(2-hydroxyethyl methacrylate)-*graft*-poly(ε-caprolactone) [P(HEMA-*g*-CL)] and prepared the electrospun NFs with gelatin. They proposed that their fibres could be used as a scaffolding biomaterial for regenerative medicine because of their biocompatibility, degradability and electrical conductivity [45]. Wang et al. suspended GO in polymers like PAN and PVP and electrospun them to form nanofiber with sandwiched GO sheets. They proposed that the interaction between CN functional group of PAN interact with a 2D layer of graphene sheets by π-π interaction after reduction of graphene oxide. In contrast, before the reduction of GO, it was governed by –CN/OH- hydrogen bond and –N- and = CO interact in the similar fashion of π-π interaction with 2D graphene sheets (Fig. 4). They obtained the conductivity of 7.5×10^3 S m^{-1} and2.5×10^3 S m^{-1} for Graphene–PAN nanofiber composite and Graphene–PVPnanofiber composite, respectively as compared to the

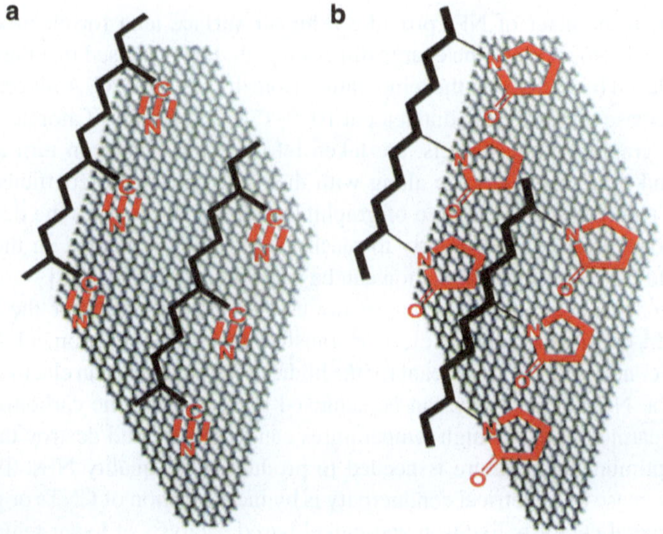

Fig. 4 Proposed interactions of graphene sheets with **a** PAN and **b** PVP (Reproduced from Ref. [46] with permission from The Royal Society of Chemistry)

pure form of PAN and PVP nanofibers with the conductivity of 2.3×10^{-11} and 4.3×10^{-7} Sm^{-1}, respectively [46].

Moayeri Ali et al. produced core–shell Polyaniline/Poly(methyl methacrylate) (PANi/PMMA) nanofibers and incorporated them with 1-pyrenebutanoic acid; succinimidyl ester functionalized graphene (G-PBASE) by using coaxial electrospinning. The produced nanofiber possess the conductivity of 30.25 S cm^{-1},which were three times more than the parent PANi nanofibers [47]. They had also doped PANi with camphor-10-sulfonic acid (HCSA), and blended it with poly(ethylene oxide) (PEO), along with the use of 1-pyrenebutanoic acid, succinimidyl ester functionalized graphene (G-PBASE) as filler. The electrical conductance for PANi/PEO/G-PBASE nanofibers was 2.41 times higher than PANi/PEO nanofibre, which was $9.92 \times 10^{-4} \pm 2.94 \times 10^{-4}$ S cm^{-1} and $2.39 \times 10^{-3} \pm 3.76 \times 10^{-5}$ S cm^{-1}, respectively [48].

6 Parameters for the Synthesis of Better Polymer Graphene-Based Nanofibers

Viscosity [49], the surface tension of the fibre precursor solution [50], applied electric field along with the distance to the collector [51], and flow rate [50] are the essential parameters that are always better to be optimized to produce the best quality NFs. Its structure and morphology can be optimized by keeping some crucial points in mind, which are the diameter of NFs, the atomic structure of NFs, surface area, pore size distribution, the particle size of metal oxides to be incorporated in NFs.

The small diameter of NFs provides a higher surface area for electrochemical reactions and also shortens the charge diffusion path. It is proposed that the diameter can be reduced by increasing the temperature from 400 to 2600 °C. A report found to have a decrease of 30–50% in diameter at 1000 °C [52]. In terms of atomic structure, degree of graphitization, defects, are taken into account, which in turn affect the sodium and lithium-ion storage along with their charge transfer coefficients. Heat treatment can increase the degree of graphitization which removes the defects and carbon atoms are arranged orderly to reach the graphitic structure. In the case of PAN-based CNFs, full graphitization can be achieved at 2000 °C [53].

Doping with heteroatoms is one of the essential means to tailor the NFs like pyridinic M-doping increase the electron transfer in the redox reaction in LABs [54]. Electrical conductivity is significant for the higher performance of an electrochemical device. The NFs conductivity can be achieved by increasing the carbonization by thermal treatment, but too high temperatures can aggregate and destroy them [55]. So, the optimum temperature is needed to produce good quality NFs. Two other ways to increase the electrical conductivity is by incorporation of CNTs or graphene by using metal catalysts like iron and nickel-based catalysts at lower temperatures [56–58]. A large surface area and optimum pore size distribution (macroporous or mesoporous) can provide higher electrochemical performance as it can offer more reaction sites, its ability to accommodate active metal oxides gets improved, maintain or buffering volume changes during an electrochemical reaction and improves the electric double-layer capacitance. Macropores are adequate for the permeation of electrolytes, whereas mesoporous distribution helps in charge transfer and works as active sites for charge storage for sodium and lithium ions [59]. And also metal oxides which possess high capacitances [55], having better dispersion qualities [60] and morphology governs the electrochemical performance of NF made electrodes.

As mentioned earlier, another way to increase the electrical conductivity of incorporated metal and metal oxide like mixing of silver, nickel, ruthenium, silicon and tin in GBEENs. Ma et al. obtained a fused film by electrospinning of polyurethane polymer solution followed by immersion in 0.05 wt% of AgNps-graphene nanosheets (GNS). They found that the film had a surface resistance of 150Ω/sq. The produced sheets were highly durable and of higher strength, they also predicted that if the ratio of AgNP-GNS solution is maintained at 3:1, the increase in the magnitude of the resistance would be less even when the film bends at 90 degrees [61]. So far, the discussion of NFs was carried out; now the following section discusses the utilization of polymer graphene NFs in energy devices.

7 Utilization of Polymer Graphene-Based Nanofibers in Lithium-Ion Batteries and Other Energy Storage Devices

Nanofibres found applications in different fields like sensors, biomedical applications, photonics, water treatment, catalysis and especially energy storage devices. If we consider the energy devices mainly, batteries are one of the essential devices which can convert stored chemical energy into electrical energy. It is made up of a metallic or plastic covering that has an anode (positive terminal), a cathode (negative terminal) along with electrolytes which help ions to travel between them. A permeable membrane called separator also presents to prevent a short circuit between the positive and negative terminal, allowing the ion charge carriers for passing to complete the circuit during current transit. When the battery is connected and the completion of the circuit takes place, oxidation occurs at the anode, where the electrolytes combine with the anode and knock out the electron producing one compound. At the cathode, the cathode material, ions and free-electron react to form another compound. In this way, the transfer of electron/ production of current continues until the electrode runs out or the reaction medium diminishes.

Two types of batteries include primary and secondary batteries. Primary batteries are non-rechargeable, disposable because the reactions occurring in them are irreversible, e.g. alkaline batteries, Zn-carbon batteries. Secondary batteries are rechargeable because the chemicals present in these batteries can undergo reversible reactions. Examples are lead-acid batteries, lithium-ion batteries [62].

Battery capacity (ampere-hours) is the measurement of the charge stores in the battery, which depends on the presence of active material in the battery. In other terms, the maximum amount of energy of the battery that can be obtained is its battery capacity. Since graphene is of lower weight, it acts as a nanofiller and helps in the production of nanocomposites. When graphene was electrospun with different synthetic and natural polymers; the properties of the polymers such as mechanical strength, conductivity and thermal stability were improved. However, the dispersion of graphene into a polymer matrix, its alignment and appropriate loadings are significant challenges to prepare optimized graphene-based nanofibers.

Loading is the crucial step to maintain the chemical affinity and flexibility; graphene can either be used as graphene oxide sheets for loading in the polymer by in situ polymerization, by blending graphene into the solution or by melting and mixing, followed by the reduction of these nanofibers by chemical method or high temperature [63]. The thermal reduction was found to be better over chemical methods for the improvement of the conductivity of graphene oxides. It is proposed that it helps in the removal of oxygen functionalities [64]. However, the homogeneous dispersion of the graphene on polymer nanofibers cannot be obtained, which brings several challenges [65, 66]. The morphology of GO and rGO using TEM and FE-SEM is shown in Fig. 5 as an example from ref. [67].

Pan et al. prepared CNFs using PAN and then mixing PAN with GO to form G-CNFs and later doping of nitrogen into G-CNFs produced NG-CNFs A smooth outer

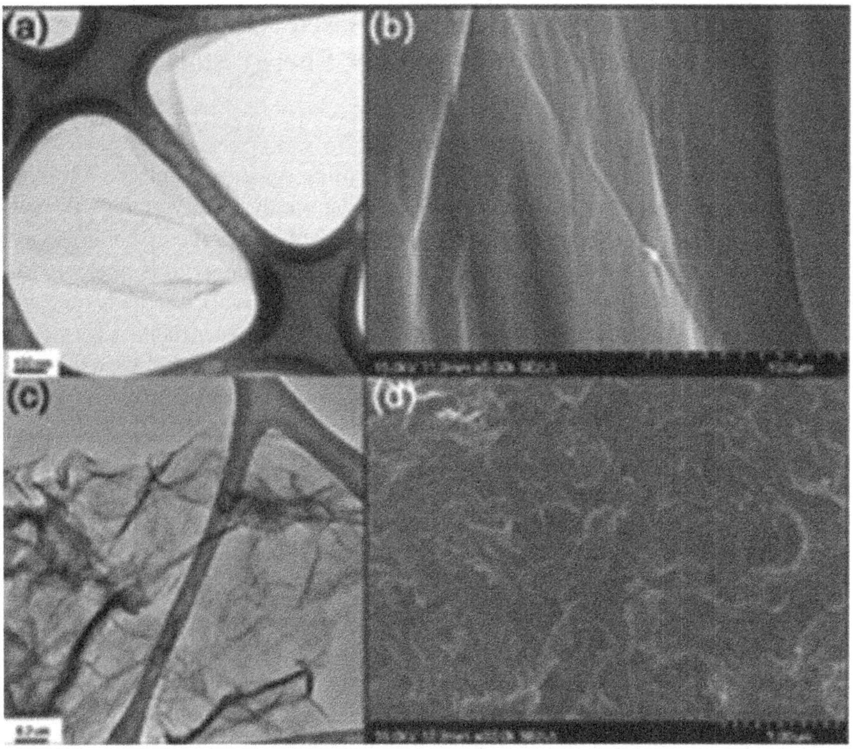

Fig. 5 TEM images of **a** GO and **c** reduced GO; FE-SEM images of **b** GO and **d** reduced GO (Reproduced from Ref. [67] with permission from The Royal Society of Chemistry)

surface with flexuous morphology of CNFs be seen in Fig. 6a and d, On fabrication with GO, the proper entanglement and uniform networks were formed in G-CNFs (Fig. 6b and e), which was not much changed on doping with nitrogen in NG-CNFs (Fig. 6c and f). Characteristic properties like specific capacitance, electrosorption capacity, the specific surface area was found to be enhanced in NG-CNFs as compared to CNFs and G-CNFs, which made an NG-CNFs a better candidate for the utilization as electrode material [63].

Liu et al. developed graphene wrapped carbon nanofiber@molybdenumdisulfide (GCNF@MoS_2) hybrids as hydrogen evolution reaction (HER) electrocatalyst by electrospinning, solution soaking and solvothermal methods. Layered graphene behaved as a "bridge" between three-dimensional (3D) carbon nanofibers and electrocatalytically active MoS_2, and helped in providing a conductive pathway that assisted in the electron transfer. They suggested that the electrical conductivity of GCNF membrane is 56.2 S m^{-1}, which is almost three times of pure CNF membrane (19.6 S m^{-1}), showing the prominence of wrapped graphene sheets [68]. Notably, Graphene and CNFs showed strong interfacial interaction made the uniform wrapping between them, provided an opportunity to anchor MoS_2nanosheets on graphene.

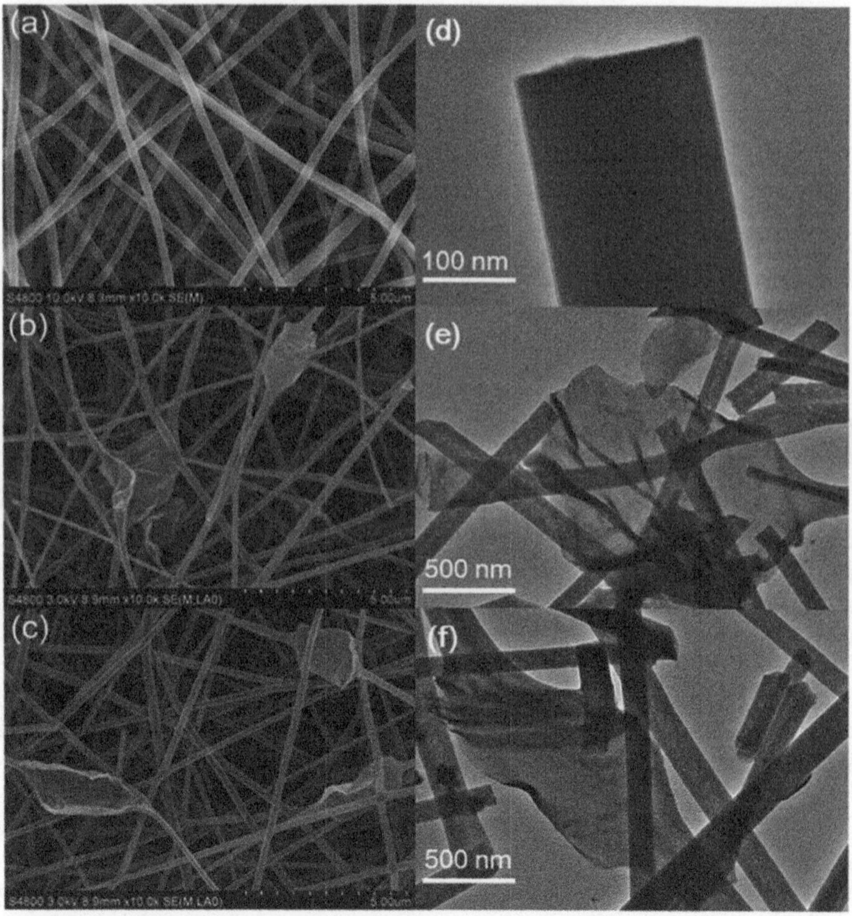

Fig. 6 FE-SEM images of **a** CNFs, **b** G–CNFs and **c** NG–CNFs / TEM images of **d** CNFs, **e** G–CNFs and **f** NG–CNFs (Reproduced from Ref. [63] with permission from The Royal Society of Chemistry)

Such a 3D template of MoS_2 grew into a morphology having a vertical orientation of small nanosheets, exposing the active sites on edges. Such impacts provided GCNF@MoS_2-20 hybrid with better performance as compared to GCNF@MoS_2-5, GCNF@MoS_2-10 and GCNF@MoS_2-40, the numbers (5,10, 20, 40) denotes the amount of $(NH_4)_2MoS_4$ in mg which was the precursor of MoS_2. Thus, they proposed that GCNF@MoS_2-20 can be utilized for the application for HER electrocatalyst.

Kim et al. prepared the porous Si/Ni/CNF composites by electrospinning and thermal treatment of silicon NPs and graphene-covered nickel particles (Fig. 7). Nickel acetate was used as the source of the Ni particles; the produced nickel particles acted as the catalyst for the generation of graphene layers by the graphitization of the PVA at the low temperature [69, 70]. The silicon NPs were completely

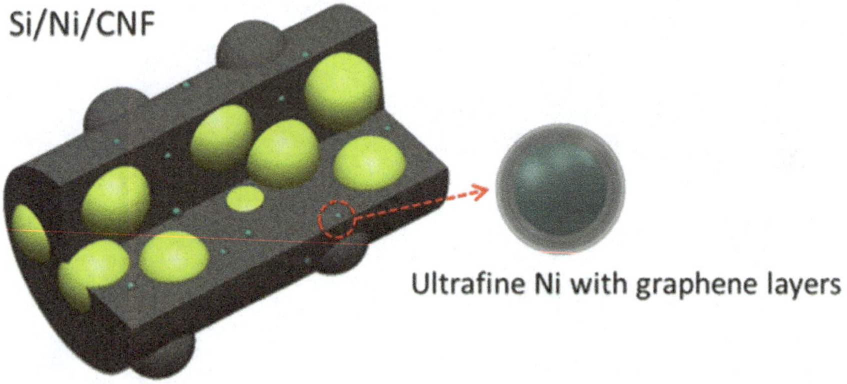

Fig. 7 Schematic demonstration of Si/Ni/CNF composite (Reproduced from Ref. [71] with permission from The Royal Society of Chemistry)

captured/encapsulated inside the CNFs. They proposed that CNFs served three purposes, one as a conductive medium for the transport of ions and electrons, second as the buffer medium for the stress-relieving arising due to the Si expansion and third, they prevented the aggregation of silicon NPs during charging/discharging. The conductivity of the Si/Ni/CNF composite was 2.31×10^{-3} and etched Si/Ni/CNF composites was 9.5×10^{-5} S cm^{-1}, which was higher in comparison to Si/CNF (1.8×10^{-5} S cm^{-1}) [71]. Apart from conductivity, Si/Ni/CNF surface area and pore size were almost double than that of Si/CNF. So they developed electrodes for LIBs by using CNFs which contained uniformly dispersed Si nanoparticles and Ni particles inside the graphene layers. They found that Si/Ni/CNF electrodes provided an outstanding reversible capacity of 1045 mAh g^{-1} even after 50 cycles at 100 mA g^{-1}, which was much higher as compared to the neat Silicon, Si/CNF or Si/CNF/Ni-etch electrodes.

Jayavelet al. synthesized graphene oxide–V_2O_5 (GVO) composite nanofibers by an electrospinning technique using vanadium acetylacetonate{$V(C_5H_7O_2)_3$}, graphene oxide and PVP (polyvinyl pyrrolidone). The morphology of V_2O_5 nanofibers were uniform and possessed a smooth surface. They proposed that the addition of graphene oxide did not affect the morphology or diameter much. But the temperature increase to 550 °C during annealing the diameter of the nanofibers were found to be shrinking, due to the PVP decomposition and onset of the crystallization. They found that the capacitance of the GVO annealed at 550 °C using CV curves [72] was 453.824 Fg^{-1} as compared to V_2O_5 electrode, which was 252.7 Fg^{-1} at a scan rate of 10 mVs^{-1}, showing a definite improvement for the supercapacitor applications [73]. Electrical conductivity and the high surface area of the graphene oxide improved the performance of the GVO. The potassium ion diffusion into the GVO pores was responsible for the lesser specific capacitance at the higher scan rate [74, 75]. So they believed that GVO could be a suitable candidate for the utilization in the electrochemical devices.

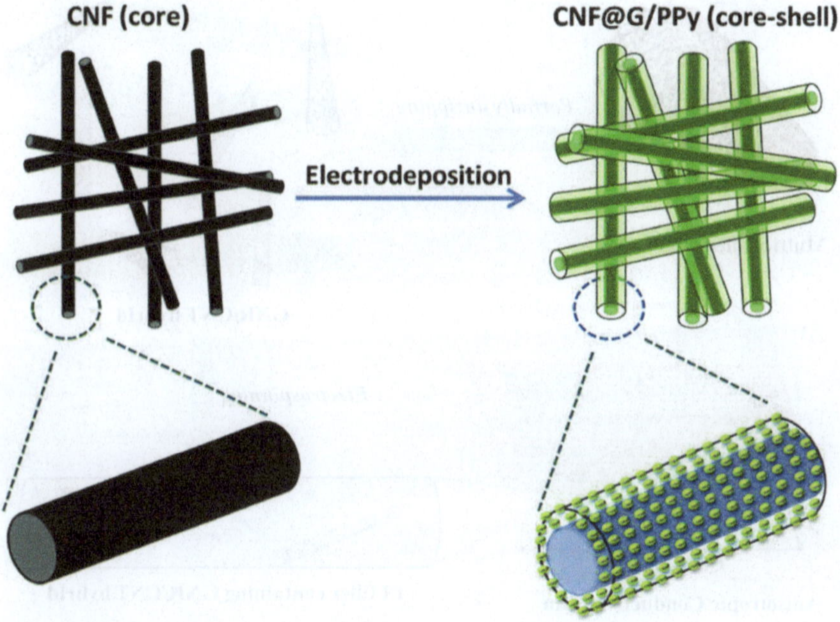

Fig. 8 Schematic illustration of the core–shell structure (Reproduced from Ref. [76] with permission from The Royal Society of Chemistry)

Pandikumar, Huang and Lim et al. fabricated graphene/polypyr role-coated carbon nanofibers (Fig. 8) using PAN, GO and pyrrole. First, they prepared CNF nanofibers by electrospinning and CNFs surface was coated with the graphene/polypyrrole (G/PPy) composite by electrode position. It was found that GO was reduced along with which pyrrole was also electro-polymerized to PPy during the electrodeposition. They proposed that the CNF were negligibly contributing to the total capacitance because the charge storage process was occurring on the surface or near-surface and CNFs were shielded by the coating of the G/PPy composite. They also prepared CNF@PPyelectrode and the specific capacitance for that electrode was 265 F g^{-1}. They showed that the charge transfer resistance was reduced because the distance for the electron transfer was decreased after the incorporation of graphene in polypyrrole chains, which leads to an increase in specific capacitance up to 386 Fg^{-1}at 2 mV s^{-1} which was almost 46% more as compared to CNF@PPy electrode, along with the improvement in the stability [76]. The concentration of the pyrrole also played an important role and they showed that when the concentration was changed from 0.2 to 0.1 and 0.3 M, the specific capacitance gradually dropped to 333 $Fg^{-1}$and 291Fg^{-1}, respectively. They also showed that at 10 mV s^{-1} the specific capacitance of CNF@G/PPy-0.2 M remained 238 Fg^{-1} with the capacity retention of 63%, but, when the CNF@PPy was studied, it was found that capacity drastically dropped to 4%. This observation showed that at higher current densities, the slow diffusion dominates during the shortening of the diffusion time in the polypyrroles [77]; in

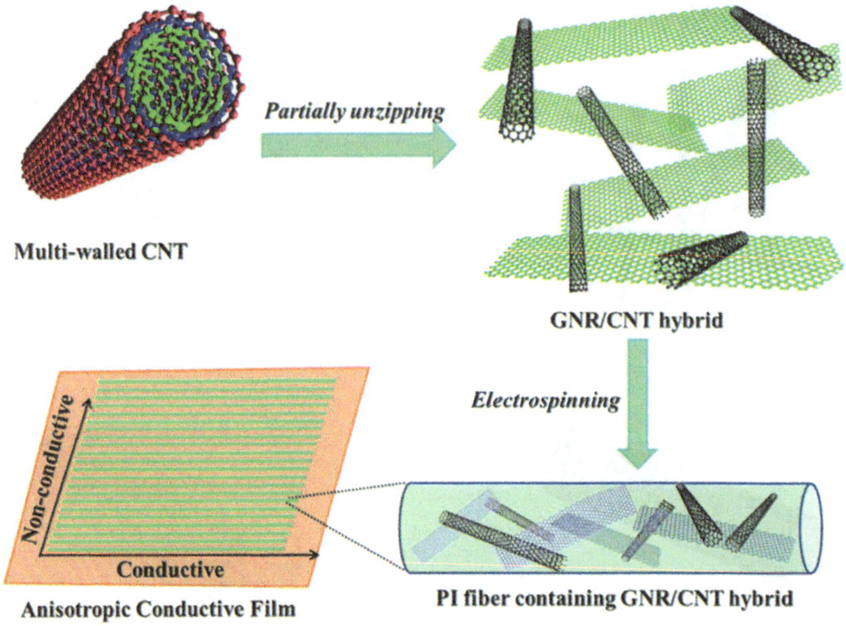

Fig. 9 Scheme showing the fabrication of PI–GNR/CNT films with GNR/CNT hybrid as a nanofiller [78]

simple language, the internal resistance gets enhanced. Whereas, the incorporation of the graphene significantly reduced the internal resistance and retained the specific capacitance in the CNF@G/PPy electrode.

Liu et al. utilized polyimide, graphene nanoribbon and carbon nanotubes and fabricated PI–GNR/CNT composite film using electrospinning. Schematic fabrication is shown in Fig. 9; in brief, it involves ultrasonication which unzips multi-walled carbon nanotubes (MWCNTs) to produce GNR sheets and residual CNTs. The residual CNTs were acting as connection bridges between different GNR sheets. Later, electrospinning with PI, they produced flexible PI–GNR/CNT films (Fig. 9). Good stability, along with a unique anisotropic conductivity, was observed with the conductivity value of 8.3×10^{-2} Sm^{-1} in the parallel direction and 7.2×10^{-8} Sm^{-1} in the perpendicular direction [78].

Liu et al. utilized a pre-oxidized electrospun PAN nanofiber membrane and soak it in an aqueous dispersion of GO, followed by the carbonization process to obtain a hybrid GCNF membrane of electrospun CNF wrapped with graphene sheets. They uniformly grew the WS_2nanosheets on the obtained GCNF with the 3D porous structure using a solvothermal method (Fig. 10). The WS_2/GCNF hybrid membrane possesses enhanced electrical conductivity and structural stability. A reversible capacity of 1128.2 mAh g^{-1} was obtained using WS_2/GCNF membrane. They proposed that such exciting results were due to the high conductive pathways provided by GCNF, along with the excellent dispersion of the WS_2nanosheets which

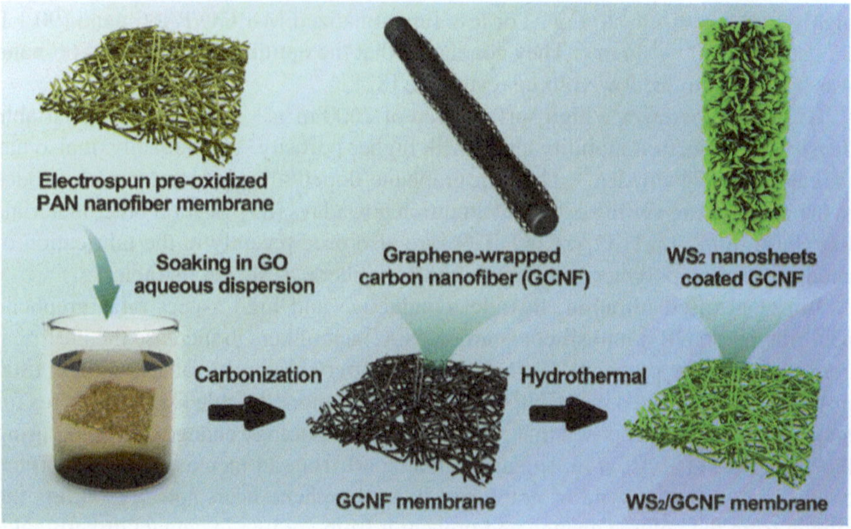

Fig. 10 Representation of the preparation method for flexible WS_2/GCNF hybrid membranes (Reproduced from Ref. [80] with permission from The Royal Society of Chemistry)

made the rapid transport of the ions easier and also made the lithium ions insertion/extraction feasible into the layered WS_2. They didn't use the binders, which also made the lithium-ion access to the active site more accessible [79]. They have also proposed that GCNF mitigated the WS_2 volumetric expansion, which resulted in excellent electrochemical cyclic stability for long term cycling. On changing the scan rates, the rate performance of the WS_2/GCNF hybrid membranes were better as compared to the bare WS_2membrane due to the rapid transfer of charges between the WS_2nanosheets and highly conductive GCNF nanofibers. They also showed the internal resistance was significantly reduced in WS_2/GCNF hybrid membranes as compared to the WS_2 membranes using Nyquist plots. They proposed that such fantastic electrochemical performance make the WS_2 /GCNF membrane a suitable candidate to be used in anode materials for lithium-ion batteries.

Pan et al. prepared nitrogen-doped reduced graphene oxide–carbon nanofiber, under ammonia atmosphere. The NG–CNF composite electrode showed a specific capacitance of 337.85 F g^{-1}), which was far more than that of pure carbon nanofibers (171.28 F g^{-1}) or the reduced graphene oxide–carbon nanofiber composite (264.32 F g^{-1}). They believed that the enhanced performance of the NG–CNF was because of the formation of a "*plane-to-line*" conducting network in the composite, assisting the electron transfer and ion transport and also increase the specific surface area due to nitrogen doping [63]. Hu et al. proposed that the dipping process incorporates rGO sheets into a polyamide-66 (PA66) and microfibers which leads to an increase in the capacitance. They studied the mass-based specific capacitance ($C_{S,M}$) [67, 81] of the functionalized medium diameter (250–450 nm) RGO nanofibers with PA-66. The M-RGO/PA66-nano was found to have a higher ($C_{S,M}$) value (280 F g^{-1}) than that of highly functionalized (20 F g^{-1}) or less functionalized M-RGO/PA66-nano (90.1 F g^{-1}) at 10 mV s^{-1} scan rate. They concluded that the optimized M-RGO/PA66-nano was better due to the low volume resistance [82].

Graphene possesses a high surface area of 2600 m^2g^{-1} or more, with a variable density, high electron mobility along with higher porosity. Theoretically, it also has the capacity of 744 mAh g^{-1}. Thus, the graphene-doped/mixed separator or electrodes in the batteries are getting excellent attention nowadays [83]. Several other materials like SnO_2, Si/C, $Li_4Ti_5O_{12}$, CoO, TiO_2 are also used recently in the fabrication of anodes based on polymer graphene-based nanofibers, as shown in Table 1.

Joo et al. used ultrathin, flexible, conductive and high aspect ratio graphene nanoribbons (GNRs) into silicon–carbon (Si/C) nanofibers to increase the electrical conductivity. The prepared GNR/Si/C fibres offered capacity of more than 1800 mAh g^{-1}. They showed that GNR/Si/C fibres were more durable than Si/C fibres for anode material and proposed that no volume or morphology change occurred during the charging [112]. Ci et al. suggested that apart from an increase in the electrical conductivity, another type of graphene/silicon graphene heart network buffers the silicon from the volume change and prevent it from coming in contact directly with the electrolyte. These composites demonstrated that high lithium storage with a high rate performance of 543 mAh g^{-1} capacity at 1000 mA g^{-1} rate [104].

Cho et al. synthesized SnO_2 hollow nanofibers (SnO_2 hNFs) with a coat of carbon and wrapping of the graphene oxide layer. These SnO_2@C@G hNFs exhibited a high lithium storage ability with 1600 mAh g^{-1}capacitance, which also maintained a value of 900 mAh g^{-1} even after 50 cycles [113]. They proposed that the excess capacity at the lower potential region was due to the decomposition of the electrolytes resulting in the organic layer formation on the SnO_2hNFs surface [114, 115]. They also proposed that the Li-ion insertion/extraction pathway through the surface of SnO_2@C@G leads to the excess capacity value and the engineering and rational design of their nanostructure and conducting framework provided the significant rate capability and cyclic stability. They suggested that in the case of SnO_2hNFs, the formation of solid electrolyte interphase (SEI) layer takes place due to the expansion and contraction of the SnO_2hNFs. This SEI layer gets thicker after repetitive 50 cycles which drastically dropped the electrochemical performance of the bare SnO_2hNFs. This SEI layer formation was prevented in the case of SnO_2@C@G, which retained a capacity of around 18% after 50 cycles, due to the coating of the graphene layer. In other words, the expansion of SnO_2hNFsat the time of the lithiation process was restricted due to the coating of the carbon layer and wrapped GOdue to high mechanical flexibility and superb electronic conductivity properties.

Recently, Cho et al. coated SnO_2 hNFs with polypyrrole layer in a nitrogen atmosphere by electrospinning followed by annealing to prepare nitrogen-doped carbon-coated SnO_2 hNFs (SnO_2/NC hNFs). The N-doped carbon layer was a layer-by-layer stack of the modified graphene that included the nitrogen element attached to the carbon lattice edges. They proposed that the N-doped carbon layer significantly enhanced the electrochemical performance due to the presence of a large number of defect sites in the N-doped carbon layer [116, 117]. The defect sites and vacancies served as the active site for lithium ions, which favoured enhanced intercalation

Table 1 Graphene-based nanofibers with their electrochemical performance

Recent polymer nanofibers with graphene	Electrochemical performance	References
Carbon nanofibers encapsulated Silicon/Graphene Oxide	The capacity of 497 mAh g^{-1}, A current density of 1 A g^{-1}	[84]
Nitrogen-doped GCNF	206 mAh g^{-1}, 200mAg^{-1}	[85]
Fe_2O_3/rGO/ CNFs film	811mAh g^{-1}, 0.1 A g^{-1}	[86]
TiO_2@PCNF-GA_2	250.1mAh g^{-1}, 0.1 A g^{-1}	[87]
C/Cu/rGO	400mAh g^{-1}, 1 A g^{-1}	[88]
Core-shell mesoporous single-crystalline MnO_xnanofibers@graphene (PSCMnO_x@G)	1072 mAh g^{-1}, 0.1 A g^{-1}	[89]
CNFs/MnO/rGO	1118 mAh g^{-1}, 0.1 A g^{-1}	[90]
Li_2FeSiO_4/CNF/rGO nanocomposites	260 mAh g^{-1}	[91]
TiO_2/GO	68mAh g^{-1}, 0.5 C	[92]
rGO /$ZnMn_2O_4$/CNFs	1142mAh g^{-1}, 100 mA g^{-1}	[93]
Cobalt Oxide/G	1591.4mAh g^{-1}	[94]
3D SnO_2/GNF	763.9mAh g^{-1}, 100 mA g^{-1}	[95]
Si/P-CNFs@rGO	1851.3mAh g^{-1}, 0.2 A g^{-1}	[96]
G/$Ca_2Ge_7O_{16}$ NFs composite	602mAh g^{-1}, 0.6 mA cm^{-2}	[97]
MnO/CNFs@G	946.5mAh g^{-1}, 0.1 A g^{-1}	[98]
$Mn_{0.5}Co_{2.5}O_4$@G	950mAh g^{-1}, 0.2 A g^{-1}	[99]
MnO/C@rGO	1148mAh g^{-1}, 0.1 A g^{-1}	[100]
rGO@TiO_2	200mAh g^{-1}, 0.1 C	[101]
$ZnFe_2O_4$-G	2166 mAh g^{-1}, 0.05 C	[102]
Ge@CNFs/3D graphene	594 mA h g^{-1}, 0.8 A g^{-1}	[103]
G/Si@CFs	543 mAh g^{-1}, 1000 mA g^{-1}	[104]
G/TiO_2	284.4 mAh g^{-1}, 100 mA g^{-1}	[105]
rGO/Fe-Fe_3C NF	558mAh g^{-1}, 1500 mA g^{-1}	[106]
Co_3O_4@rGO	900mAh g^{-1}, 1 A g^{-1}	[107]
Si@RGO@C NFs	1228mAh g^{-1}	[108]
G@C/Fe_3O_4 NFs	872mAh g^{-1}, 100 mA g^{-1}	[109]
Thermally reduced GO/CNFs-Sn/SnO_2	600mAh g^{-1}, 0.2 C	[110]
Co_3O_4NFs@rGO	615mAh g^{-1}	[111]

Note Please check the references to know the number of cycles and stability

of Li^+ ions [118]. They proposed that the vacancies could also trap the lithium ions at the first cycle, which could get extended in the following cycles by which it favoured the lithium-ion insertion. Thus, the nitrogen-doped carbon shell layers boosted the capacity to 1648 mAh g^{-1}, which was 427% higher than neat SnO_2 hNFs (386 mAh g^{-1}) [119].

Interestingly, a micron-sized cotton fibre when covered by graphene/Co-doped SnO_2(CGN/SnO_2-Co) composite, the phase structure of SnO_2 remained unaffected. In contrast, it also gave a high rate of discharge capacity of 549.6 mAh g^{-1}[120]. Lu et al. prepared 1D graphene/TiO_2 composite nanofibers in which graphene prevents the agglomeration of TiO_2anatase and also maintained a high surface area. The produced graphene/TiO_2 nanofibers possess the high reversible capacity of 284.4 mAh g^{-1} at a current density of 100 mA g^{-1} after 100 cycles [105]. Recently, Fedkiw et al. thermally reduced GO-containing CNFs and house Sn/SnO_2. They noticed that the oxidation state of tin in TRGO/CNFs was partly temperature-dependent. And fibres with $\leq$ 10 wt% Sn (IV) produced the smaller size particles (<15 nm diameter), which also reduced absolute volume expansions and lead to higher capacities of 600 mAh g^{-1}[110]. Similarly, a 3D SnO_2/GNF served as a binder-free anode possessing excellent cycling stability, showing a high capacity of 763.9 mAh g^{-1} after 300 cycles at 100 mA g^{-1} current density, synergistically [95]. Severe particle pulverization was also minimized in silicon electrodes by encapsulating them in PAN carbon nanofiber in the conjugation of graphene oxide layers, forming a 3D network in Li-ion batteries [84]. The higher electrochemical performance was achieved by Nair et al. by utilizing Li_2FeSiO_4/CNF/rGO nanocomposites which were synthesized by CNFs and rGO. They believed that a large contact area between the electrolyte and the cathode was obtained for efficient insertion and extraction of Li-ions by using these porous nanocomposites [91].

Xie et al. utilized an aerogel of porous $Li_4Ti_5O_{12}$ (PLTO) nanofibers with 3D-interconnected graphene framework (G-PLTO), they fabricated it for sodium storage in sodium-ion batteries possessing the capacity of 195 mAh g^{-1}[121]. Kim et al. designed a bifunctional catalyst made up of electrospun 1D Co_3O_4 nanofibers (NFs) covered both sides with the 2D graphene nanoflakes (GNFs) for an oxygen electrode in Li–O_2 batteries. The obtained Co_3O_4 NF/GNF showed a high discharge capacity of 10,500 mAh g^{-1}[122].

Yang et al. used electrospinning in combination with the electrospraying to obtain [MnO/CNFs@rGO, (MCG)] composite film electrodes by using a solution of PAN in DMF dispersed with manganese oxide and a solution of GO in ethanol followed by its carbonization (Fig. 11). They noticed a high reversible capacity of 1148 mAh g^{-1} at a current density of 0.1 A g^{-1} which also retained its value till 332 mAh g^{-1} at a current density of 5 A g^{-1} even after 4500 cycles [100]. It was interesting to note that the capacity of their MCG-800 gradually gets increased in the initial 1500 cycles. They proposed that more Mn^{+2} get oxidized to Mn^{+3}, thus increasing the capacity. Their MCG nanowire's rough surface also provided a shorter distance for the transportation of the ions and charges as well as exposed more active sites to store them which were also proposed as another reason for the enhanced capacity. They suggested that the interparticle contact resistance proved the high structural stability or integrity of MCG-800 from the Nyquist plot calculation at different cycles.

Chen et al. fabricated flexible MnO/carbon nanofibers (MnO/CNFs@G) membrane wrapped with graphene using electrospinning of Manganese acetate and PAN in DMF solution followed by an ambient pressure CVD process (Fig. 12) [98].

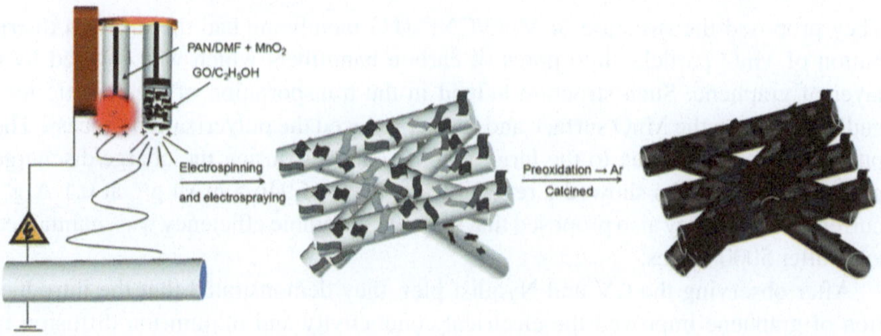

Fig. 11 Schematic representation of the preparation of MCG architecture(Reproduced from Ref. [100] with permission from The Royal Society of Chemistry)

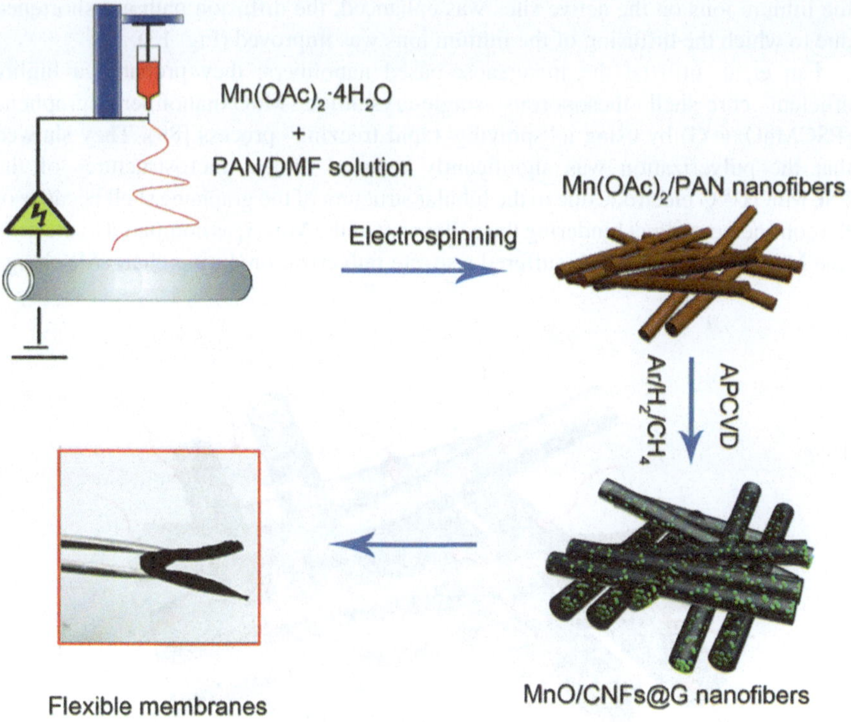

Fig. 12 Schematic representation of the fabrication of the MnO/CNFs@G nanofiber membranes (Reproduced from Ref. [98] with permission from The Royal Society of Chemistry)

They proposed the structure of MnO/CNFs@G membrane had the uniform distribution of MnO particles into pores of carbon nanofibers which was covered by a layer of graphene. Such structure helped in the transportation of electrolytic ions and electrons on the MnO surface and further reduced the pulverization process. The pulverization occurs due to the large volume changes during the charge/discharge process. Their results showed a reversible capacity of 946.5 mAh g^{-1} at 0.1 A g^{-1} current density. They also proposed that 99% of coulombic efficiency was maintained even after 5000 cycles.

After observing the CV and Nyquist plot, they demonstrated that the introduction of graphene improved the electrical conductivity and lithium-ion diffusion in MnO/CNFs. Lower impedance was attributed to the larger surface area, uniform distribution of MnO particles and the porous structure of the membrane was associated with the excellent wettability of electrolytes. Due to which the accessibility of the lithium ions on the active sites was enhanced, the diffusion path got shortened due to which the diffusion of the lithium ions was improved (Fig. 13).

Fan et al. utilized the manganese-based nanofibers; they prepared a highly efficient core-shell mesoporous single-crystalline MnOxnanofibers@graphene (PSCMnOx@G) by using a “spraying–rapid freezing” process [89]. They showed that the pulverization was significantly reduced in the microstructures of the PSCMnOx@G electrode due to the tubular structure of the graphene shell because of the confinement effect hindering the collapsing of the MnOx, as compared to the pristine MnO_2 electrode which suffered a severe pulverization during charge/discharge

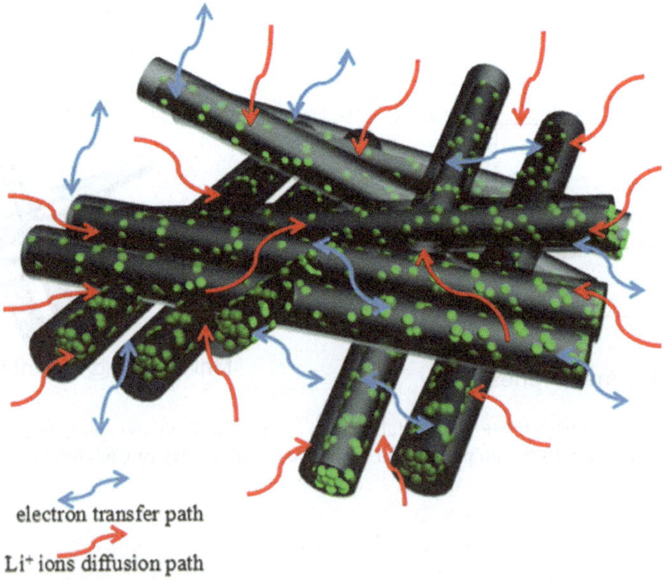

Fig. 13 Scheme showing the diffusion of lithium ions and electrons in the MnO/CNFs@G membranes (Reproduced from Ref. [98] with permission from The Royal Society of Chemistry)

processes [123]. The graphene further improved the electrical conductivity of the electrode. The mesoporous structure of the MnOx nanofiber is favourable for lithium-ion diffusion [124]. They proposed that their crystalline structure provided straight, open and short channels for lithium extraction/insertion responsible for the enhancement of the ion transfer kinetics [125, 126]. And also the Mn^{4+}–O–Mn^{3+} formation increased the conductivity of MnOx [127, 128]. They explained that the high-energy electron from the eg antibonding orbital of Mn^{3+} could move into the empty eg orbital of Mn^{4+} via. O^{2-} which can change the oxidation state of Mn^{+4} to Mn^{+3}.They proposed that due to Jahn–Teller distortion, resultant Mn^{+3} generates unstable octahedral structure with four short Mn^{4+}–O–Mn^{3+} bonds and two long Mn–O bonds, which increases the conductivity.

In short, mesopores were helping in fast lithium-ion transport along with the vertically aligned crystalline nature of the PSCMnOx@G. Double exchange interaction of manganese mixed-valence ions by the formation of Mn^{4+}–O–Mn^{3+} led to further improvement of the conductivity of MnOx (Fig. 14) [127, 129]. TheirPSCMnOx@G

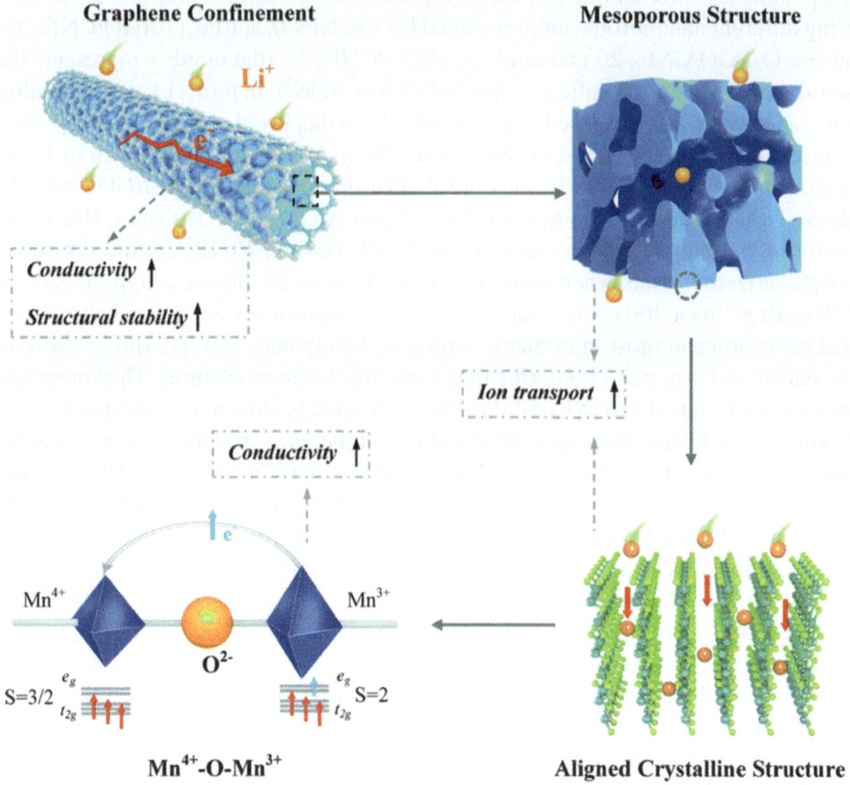

Fig. 14 The figure represents the effect of the microstructures on the structural stability, conductivity and ion transport of the $PSCMnO_x$@G (Reproduced from Ref. [89] with permission from The Royal Society of Chemistry)

showed an ultrahigh performance (1072 mAh g^{-1} at 0.1 A g^{-1} and 419 mAh g^{-1} at 10 A g^{-1}), along with excellent cycling stability (1162 mAh g^{-1} at 2 A g^{-1} over 500 cycles).

Moreover, the role of iron and cobalt oxides in conjunction with nanofibers is also apparent in lithium-ion batteries. Cui et al. prepared porous graphene-doped carbon/Fe_3O_4(GN@C/Fe_3O_4) nanofiber using polyacrylonitrile (PAN)/polymethyl methacrylate (PMMA) solution having ferric acetylacetonate and GO sheets by in situ electrospinning, followed by the thermal treatment. They found the reversible capacity of 872 mAh g^{-1} at 100 mAg^{-1} current density after 100 cycles [109]. Yoon et al. utilized a supersonic technique to synthesize rGO loaded Fe-Fe_3C composite nanofiber. They revealed rapid and simultaneous coupling of exfoliated rGO with the solution-blown nanofibers could be achieved by supersonic blowing, unlike electrospinning. The electrodes also showed a reversible capacity of 558 mAh g^{-1}at 1500 mA g^{-1} at the 200th cycle [106].

Wu et al. utilized$FeSO_4 7H_2O$, GO and a mixture of PAN and PVP as the precursors and synthesized button-shaped amorphous Fe_2O_3/rGO/CNFs-20 films by in situ electrospinning followed by one-step carbonization. They performed their studies considering different nanofibrous samples named FeO_x/CNFs-0, am-Fe_2O_3/rGO/CNFs-10, am-Fe_2O_3/rGO/CNFs-20 and am-Fe_2O_3/rGO/CNFs-50 (the number represents the amount of GO in mg, 0 indicates that no GO was added). In terms of cyclic stability and capacity, it was proposed that FeOx/CNFs-0 displayed a stable cyclic performance and a capacity of 524 mAhg^{-1} over 350 cycles due to the presence of FeO_X (particularly Fe_2O_3) active species [130]. The produced am-Fe_2O_3/rGO/CNFs-10 showed a slow capacity rise with a value of 646 mAhg^{-1} after 350 cycles. Due to the uniform distribution and smaller size, am-Fe_2O_3/rGO/CNFs-10 also did not suffer the pulverization issue, whereas am-Fe_2O_3/rGO/CNFs-20 showed a high capacity of 739 mAh g^{-1} over 400 cycles, due to its larger nanoparticle size, more electrochemical activation and most importantly a reversible polymeric gel-like film growth on the pulverized nanoparticles during the charge/discharge procedures. They proposed that the am-Fe_2O_3/rGO/CNFs-50 showed an unstable behaviour of the capacity (548 mAhg^{-1} after 350 cycles) due to its unstable protuberance structure. It is noteworthy that am-Fe_2O_3/rGO/CNFs-20 delivered a capacity of 1378 and 825 mAh g^{-1} in the first discharge and charge cycle, respectively, but the coulombic efficiency was only 60%. They proposed that the drawback was due to the irreversible SEI film formation and decomposition of the electrolytes caused by the enormous active sites on CNFs [131, 132]. Later, the electrode showed a stable reversible capacity of 811 mAh g^{-1}after 150 cycles, along with the coulombic efficiency of 100% with remarkable cyclic stability. These amorphous Fe_2O_3/rGO/CNFs-20 electrodes also possess flexibility; with such improved properties, they can be utilized to form flexible lithium-ion batteries [86].

Hu et al. used the hydrothermal method to develop cobalt oxide/G composite electrode, which showed a high reversible capacity of 1591.4 mAh g^{-1}, which remained up to 550 mAh g^{-1} after 70 cycles [94]. They constructed a cell using Co_3O_4@rGO as anode and a $LiMn_2O_4$ as a cathode which delivered excellent capacity retention with an operation voltage of around 2.0 V [107]. These Co_3O_4@rGO electrodes displayed a high coulombic efficiency, higher cyclic stability, along with a rate capability of 900 $mAhg^{-1}$ at 1A g^{-1}[107]. When utilized with manganese, using $Mn_{0.5}Co_{2.5}O_4$ as nanofiber to form $Mn_{0.5}Co_{2.5}O_4$@G composite, it showed a value of 950 mAh g^{-1} after 100 cycles at a current density of 0.2 A g^{-1}.

8 Conclusions and Perspectives

Excellent performance from batteries can be obtained using graphene with nanofibers. The notable point was excellent electrical conductivity, higher surface area, also electrospinning which is an easy fabrication process that can reduce the diameters of nanofibers up to 10 nm, having low fabrication cost with higher electrochemical performance. In short, the chapter tells that efficient electron transport and ease of lithium-ion to and fro transportation, architectural stability by graphene covering, lithium storage along with cyclic stability are the major factors to keep track of to obtain a highly efficient Lithium-ion battery. Although there are several studies done to understand and control morphologies and structure of nanofibers, still there is a need for further studies in this area.

Acknowledgements AS would like to acknowledge the Ministry of Human Resource Development (MHRD), the Council of Scientific Industrial and Research (CSIR) and IIT Indore.SM would like to acknowledge the Science & Engineering Research Board (SERB) for the National Post-Doctoral Fellowship (NPDF). SKS would like to thank the CSIR for the Senior Research Fellowship.

References

1. Bruce, P.G., Freunberger, S.A., Hardwick, L.J., Tarascon, J.-M.: Li–O_2 and Li–S batteries with high energy storage. Nat. Mater. **11**, 19–29 (2012)
2. Goodenough, J.B.: Electrochemical energy storage in a sustainable modern society. Energy Environ. Sci. **7**, 14–18 (2014)
3. Frackowiak, E., Béguin, F.: Carbon materials for the electrochemical storage of energy in capacitors. Carbon **39**, 937–950 (2001)
4. Mitchell, R.R., Gallant, B.M., Thompson, C.V., Shao-Horn, Y.: All-carbon-nanofiber electrodes for high-energy rechargeable Li–O_2 batteries. Energy Environ. Sci. **4**, 2952–2958 (2011)
5. Li, X., Kang, F., Bai, X., Shen, W.: A novel network composite cathode of $LiFePO_4$/multiwalled carbon nanotubes with high rate capability for lithium ion batteries. Electrochem. Commun. **9**, 663–666 (2007)
6. Zhang, B., Yu, Y., Liu, Y., Huang, Z.-D., He, Y.-B., Kim, J.-K.: Percolation threshold of graphene nanosheets as conductive additives in $Li_4Ti_5O_{12}$ anodes of Li-ion batteries. Nanoscale **5**, 2100–2106 (2013)
7. Paek, S.-M., Yoo, E., Honma, I.: Enhanced cyclic performance and lithium storage capacity of SnO_2/graphene nanoporous electrodes with three-dimensionally delaminated flexible structure. Nano Lett. **9**, 72–75 (2009)

8. Zheng, G., Yang, Y., Cha, J.J., Hong, S.S., Cui, Y.: Hollow carbon nanofiber-encapsulated sulfur cathodes for high specific capacity rechargeable lithium batteries. Nano. Lett. **11**, 4462–4467 (2011)
9. Wen, Z., Wang, Q., Zhang, Q., Li, J.: In Situ growth of mesoporous SnO_2 on multiwalled carbon nanotubes: a novel composite with porous-tube structure as anode for lithium batteries. Adv. Funct. Mater. **17**, 2772–2778 (2007)
10. Soldano, C., Mahmood, A., Dujardin, E.: Production, properties and potential of graphene. Carbon **48**, 2127–2150 (2010)
11. Novoselov, K.S., Fal′ko, V.I., Colombo, L., Gellert, P.R., Schwab, M.G., Kim, K.: A roadmap for graphene. Nature **490**, 192–200 (2012)
12. Yang, Y., Asiri, A.M., Tang, Z., Du, D., Lin, Y.: Graphene based materials for biomedical applications. Mater. Today. **16**, 365–373 (2013)
13. Raccichini, R., Varzi, A., Passerini, S., Scrosati, B.: The role of graphene for electrochemical energy storage. Nat. Mater. **14**, 271–279 (2015)
14. Jang, H., Park, Y.J., Chen, X., Das, T., Kim, M.-S., Ahn, J.-H.: Graphene-based flexible and stretchable electronics. Adv. Mater. **28**, 4184–4202 (2016)
15. Winter, M., Besenhard, J.O., Spahr, M.E., Novák, P.: Insertion electrode materials for rechargeable lithium batteries. Adv. Mater. **10**, 725–763 (1998)
16. Zhang, W.-J.: A review of the electrochemical performance of alloy anodes for Lithium-ion batteries. J. Power Sources. **196**, 13–24 (2011)
17. Alghoraibi, I., Alomari, S.: Different methods for nanofiber design and fabrication. In: Barhoum, A.; Bechelany, M.; Makhlouf, A., (eds.) Handbook of Nanofibers, vol. pp. 1–46. Springer International Publishing, Cham (2018)
18. Mohammadi, A., Hasan, M.A., Liedberg, B., Lundström, I., Salaneck, W.R.: Chemical vapour deposition (CVD) of conducting polymers: polypyrrole. Synth. Met. **14**, 189–197 (1986)
19. De Jong, K.P., Geus, J.W.: Carbon nanofibers: catalytic synthesis and applications. Catal. Rev. **42**, 481–510 (2000)
20. Martin-Gullon, I., Vera, J., Conesa, J.A., González, J.L., Merino, C.: Differences between carbon nanofibers produced using Fe and Ni catalysts in a floating catalyst reactor. Carbon **44**, 1572–1580 (2006)
21. Zhang, B., Liu, Y., Huang, Z., Oh, S., Yu, Y., Mai, Y.-W., Kim, J.-K.: Urchin-like $Li_4Ti_5O_{12}$–carbon nanofiber composites for high rate performance anodes in Li-ion batteries. J. Mater. Chem. **22**, 12133–12140 (2012)
22. Szosland, L., East, G.C.: The dry spinning of dibutyrylchitin fibers. J. Appl. Polym. Sci. **58**, 2459–2466 (1995)
23. Zhang, X., Liu, T., Sreekumar, T.V., Kumar, S., Hu, X., Smith, K.: Gel spinning of PVA/SWNT composite fiber. Polymer **45**, 8801–8807 (2004)
24. Dong, H.: Hessian equations with elementary symmetric functions. Commun. Partial Differential Equ. **31**, 1005–1025 (2006)
25. Wang, Y.-X., Wang, C.-G., Bai, Y.-J., Bo, Z.: Effect of the drawing process on the wet spinning of polyacrylonitrile fibers in a system of dimethyl sulfoxide and water. J. Appl. Polym. Sci. **104**, 1026–1037 (2007)
26. Minus, M., Kumar, S.: The processing, properties, and structure of carbon fibers. JOM **57**, 52–58 (2005)
27. Shen, L., Che, Q., Li, H., Zhang, X.: Mesoporous $NiCo_2O_4$ nanowire arrays grown on carbon textiles as binder-free flexible electrodes for energy storage. Adv. Funct. Mater. **24**, 2630–2637 (2014)
28. Xianwen, M., Hatton, T.A., Gregory, C.R.: A Review of Electrospun carbon fibers as electrode materials for energy storage. Curr. Org. Chem. **17**, 1390–1401 (2013)
29. Bhattacharyya, I., Molaro, M.C., Braatz, R.D., Rutledge, G.C.: Free surface electrospinning of aqueous polymer solutions from a wire electrode. Chem. Eng. J. **289**, 203–211 (2016)
30. Balaji, A., Vellayappan, M.V., John, A.A., Subramanian, A.P., Jaganathan, S.K., Supriyanto, E., Razak, S.I.A.: An insight on electrospun-nanofibers-inspired modern drug delivery system in the treatment of deadly cancers. RSC Adv. **5**, 57984–58004 (2015)

31. Yu, Y., Gu, L., Wang, C., Dhanabalan, A., van Aken, P.A., Maier, J.: Encapsulation of Sn@carbon nanoparticles in bamboo-like hollow carbon nanofibers as an anode material in lithium-based batteries. Angew. Chem. Int. Ed. **48**, 6485–6489 (2009)
32. Hwang, T.H., Lee, Y.M., Kong, B.-S., Seo, J.-S., Choi, J.W.: Electrospun core-shell fibers for robust silicon nanoparticle-based lithium ion battery anodes. Nano. Lett. **12**, 802–807 (2012)
33. Nikmaram, N., Roohinejad, S., Hashemi, S., Koubaa, M., Barba, F.J., Abbaspourrad, A., Greiner, R.: Emulsion-based systems for fabrication of electrospun nanofibers: food, pharmaceutical and biomedical applications. RSC Adv. **7**, 28951–28964 (2017)
34. Zamani, M., Prabhakaran, M.P., Ramakrishna, S.: Advances in drug delivery via electrospun and electrosprayed nanomaterials. Int. J. Nanomed. **8**, 2997–3017 (2013)
35. Liao, I.C., Chen, S., Liu, J.B., Leong, K.W.: Sustained viral gene delivery through core-shell fibers. J. Controll. Release. **139**, 48–55 (2009)
36. Yarin, A.L.: Coaxial electrospinning and emulsion electrospinning of core–shell fibers. Polym. Adv. Technol. **22**, 310–317 (2011)
37. Yang, Y., Li, X., Qi, M., Zhou, S., Weng, J.: Release pattern and structural integrity of lysozyme encapsulated in core–sheath structured poly(dl-lactide) ultrafine fibers prepared by emulsion electrospinning. Eur. J. Pharm. Biopharm. **69**, 106–116 (2008)
38. Xu, X., Yang, L., Xu, X., Wang, X., Chen, X., Liang, Q., Zeng, J., Jing, X.: Ultrafine medicated fibers electrospun from W/O emulsions. J. Controll. Release. **108**, 33–42 (2005)
39. Pelipenko, J., Kocbek, P., Kristl, J.: Critical attributes of nanofibers: preparation, drug loading, and tissue regeneration. Int. J. Pharm. **484**, 57–74 (2015)
40. Rahaman, M.S.A., Ismail, A.F., Mustafa, A.: A review of heat treatment on polyacrylonitrile fiber. Polym. Degrad. Stab. **92**, 1421–1432 (2007)
41. Jandas, P.J., Luo, J., Prabakaran, K., Chen, F., Fu, Y. Q.: Highly stable, love-mode surface acoustic wave biosensor using Au nanoparticle-MoS_2-rGO nano-cluster doped polyimide nanocomposite for the selective detection of carcinoembryonic antigen. Mater Chem Phys. **246**, 122800 (2020)
42. Ua-amnueychai, W., Kodama, S., Tanthapanichakoon, W., Sekiguchi, H.: Preparation of zinc coated PMMA using solid precursor by gliding arc discharge. Chem. Eng. J. **278**, 301–308 (2015)
43. Nisticò, R., Magnacca, G., Jadhav, S.A., Scalarone, D.: Polystyrene-block-poly(ethylene oxide) copolymers as templates for stacked, spherical large-mesopore silica coatings: dependence of silica pore size on the PS/PEO ratio. Beilstein J. Nanotechnol. **7**, 1454–1460 (2016)
44. Matsumoto, H., Imaizumi, S., Konosu, Y., Ashizawa, M., Minagawa, M., Tanioka, A., Lu, W., Tour, J.M.: Electrospun composite nanofiber yarns containing oriented graphene nanoribbons. ACS Appl. Mater. Interf. **5**, 6225–6231 (2013)
45. Massoumi, B., Ghandomi, F., Abbasian, M., Eskandani, M., Jaymand, M.: Surface functionalization of graphene oxide with poly(2-hydroxyethyl methacrylate)-graft-poly(ε-caprolactone) and its electrospun nanofibers with gelatin. Appl. Phys. A. **122**, 1000 (2016)
46. Wang, Y., Tang, J., Xie, S., Liu, J., Xin, Z., Liu, X., Belfiore, L.A.: Leveling graphene sheets through electrospinning and their conductivity. RSC Adv. **5**, 42174–42177 (2015)
47. Moayeri, A., Ajji, A.: Core shell structured graphene filled polyaniline/poly(methyl methacrylate) nanofibers by coaxial electrospinning. Nanosci. Nanotechnol. Lett. **8**, 129–134 (2016)
48. Moayeri, A., Ajji, A.: Fabrication of polyaniline/poly(ethylene oxide)/non-covalently functionalized graphene nanofibers via electrospinning. Synth. Met. **200**, 7–15 (2015)
49. Uyar, T., Besenbacher, F.: Electrospinning of uniform polystyrene fibers: the effect of solvent conductivity. Polymer **49**, 5336–5343 (2008)
50. Wang, C., Hsu, C.-H., Lin, J.-H.: Scaling laws in electrospinning of polystyrene solutions. Macromolecules **39**, 7662–7672 (2006)
51. Fang, J., Wang, H., Niu, H., Lin, T., Wang, X.: Evolution of fiber morphology during electrospinning. J. Appl. Polym. Sci. **118**, 2553–2561 (2010)
52. Mittal, J., Bahl, O.P., Mathur, R.B.: Single step carbonization and graphitization of highly stabilized PAN fibers. Carbon **35**, 1196–1197 (1997)

53. Yan, H., Mahanta, N.K., Wang, B., Wang, S., Abramson, A.R., Cakmak, M.: Structural evolution in graphitization of nanofibers and mats from electrospun polyimide–mesophase pitch blends. Carbon **71**, 303–318 (2014)
54. Li, Q., Xu, P., Gao, W., Ma, S., Zhang, G., Cao, R., Cho, J., Wang, H.-L., Wu, G.: Graphene/graphene-tube nanocomposites templated from cage-containing metal-organic frameworks for oxygen reduction in Li–O_2 batteries. Adv. Mater. **26**, 1378–1386 (2014)
55. Zhang, B., Yu, Y., Huang, Z., He, Y.-B., Jang, D., Yoon, W.-S., Mai, Y.-W., Kang, F., Kim, J.-K.: Exceptional electrochemical performance of freestanding electrospun carbon nanofiber anodes containing ultrafine SnOx particles. Energy Environ. Sci. **5**, 9895–9902 (2012)
56. Zhang, B., Xu, Z.-L., He, Y.-B., Abouali, S., Akbari Garakani, M., Kamali Heidari, E., Kang, F., Kim, J.-K.: Exceptional rate performance of functionalized carbon nanofiber anodes containing nanopores created by (Fe) sacrificial catalyst. Nano Energy **4**, 88–96 (2014)
57. Chen, Y., Lu, Z., Zhou, L., Mai, Y.-W., Huang, H.: Triple-coaxial electrospun amorphous carbon nanotubes with hollow graphitic carbon nanospheres for high-performance Li ion batteries. Energy Environ. Sci. **5**, 7898–7902 (2012)
58. Tai, Z., Yan, X., Lang, J., Xue, Q.: Enhancement of capacitance performance of flexible carbon nanofiber paper by adding graphene nanosheets. J. Power Sources. **199**, 373–378 (2012)
59. Su, D.S., Schlögl, R.: Nanostructured carbon and carbon nanocomposites for electrochemical energy storage applications. Chemsuschem **3**, 136–168 (2010)
60. Zhu, C., Mu, X., van Aken, P.A., Yu, Y., Maier, J.: Single-layered ultrasmall nanoplates of MoS_2 embedded in carbon nanofibers with excellent electrochemical performance for lithium and sodium storage. Angew. Chem. Int. Ed. **53**, 2152–2156 (2014)
61. Huang, Y.-L., Baji, A., Tien, H.-W., Yang, Y.-K., Yang, S.-Y., Wu, S.-Y., Ma, C.-C.M., Liu, H.-Y., Mai, Y.-W., Wang, N.-H.: Self-assembly of silver–graphene hybrid on electrospun polyurethane nanofibers as flexible transparent conductive thin films. Carbon **50**, 3473–3481 (2012)
62. Graphene Batteries: Introduction And Market News | Graphene-Info (2020)
63. Liu, Y., Xu, X., Lu, T., Sun, Z., Chua, D.H.C., Pan, L.: Nitrogen-doped electrospun reduced graphene oxide–carbon nanofiber composite for capacitive deionization. RSC Adv. **5**, 34117–34124 (2015)
64. Wang, G., Dong, Q., Wu, T., Zhan, F., Zhou, M., Qiu, J.: Ultrasound-assisted preparation of electrospun carbon fiber/graphene electrodes for capacitive deionization: importance and unique role of electrical conductivity. Carbon **103**, 311–317 (2016)
65. Park, G., Kim, S., Chae, S., Han, H., Le, T.-H., Yang, K.S., Chang, M., Kim, H., Yoon, H.: Combining SWNT and graphene in polymer nanofibers: a route to unique carbon precursors for electrochemical capacitor electrodes. Langmuir **35**, 3077–3086 (2019)
66. Ege, D., Kamali, A.R., Boccaccini, A.R.: Graphene oxide/polymer-based biomaterials. Adv. Eng. Mater. **19**, 1700627 (2017)
67. Yang, S.-Y., Chang, K.-H., Tien, H.-W., Lee, Y.-F., Li, S.-M., Wang, Y.-S., Wang, J.-Y., Ma, C.-C.M., Hu, C.-C.: Design and tailoring of a hierarchical graphene-carbon nanotube architecture for supercapacitors. J. Mater. Chem. **21**, 2374–2380 (2011)
68. Gu, H., Huang, Y., Zuo, L., Fan, W., Liu, T.: Graphene sheets wrapped carbon nanofibers as a highly conductive three-dimensional framework for perpendicularly anchoring of MoS_2: advanced electrocatalysts for hydrogen evolution reaction. Electrochim. Acta. **219**, 604–613 (2016)
69. Chen, Y., Li, X., Park, K., Song, J., Hong, J., Zhou, L., Mai, Y.-W., Huang, H., Goodenough, J.B.: Hollow carbon-nanotube/carbon-nanofiber hybrid anodes for li-ion batteries. J. Am. Chem. Soc. **135**, 16280–16283 (2013)
70. Helveg, S., López-Cartes, C., Sehested, J., Hansen, P.L., Clausen, B.S., Rostrup-Nielsen, J.R., Abild-Pedersen, F., Nørskov, J.K.: Atomic-scale imaging of carbon nanofibre growth. Nature **427**, 426–429 (2004)
71. Xu, Z.-L., Zhang, B., Zhou, Z.-Q., Abouali, S., Akbari Garakani, M., Huang, J., Huang, J.-Q., Kim, J.-K.: Carbon nanofibers containing Si nanoparticles and graphene-covered Ni for high performance anodes in Li ion batteries. RSC Adv. **4**, 22359–22366 (2014)

72. Meher, S.K., Rao, G.R.: Ultralayered Co_3O_4 for high-performance supercapacitor applications. J. Phys. Chem. C. **115**, 15646–15654 (2011)
73. Thangappan, R., Kalaiselvam, S., Elayaperumal, A., Jayavel, R.: Synthesis of graphene oxide/vanadium pentoxide composite nanofibers by electrospinning for supercapacitor applications. Solid State Ionics **268**, 321–325 (2014)
74. Reddy, R.N., Reddy, R.G.: Porous structured vanadium oxide electrode material for electrochemical capacitors. J. Power Sources. **156**, 700–704 (2006)
75. Reddy, R.N., Reddy, R.G.: Sol–gel MnO_2 as an electrode material for electrochemical capacitors. J. Power Sources. **124**, 330–337 (2003)
76. Gan, J.K., Lim, Y.S., Pandikumar, A., Huang, N.M., Lim, H.N.: Graphene/polypyrrole-coated carbon nanofiber core–shell architecture electrode for electrochemical capacitors. RSC Adv. **5**, 12692–12699 (2015)
77. Snook, G.A., Kao, P., Best, A.S.: Conducting-polymer-based supercapacitor devices and electrodes. J. Power Sources. **196**, 1–12 (2011)
78. Liu, M., Du, Y., Miao, Y.-E., Ding, Q., He, S., Tjiu, W.W., Pan, J., Liu, T.: Anisotropic conductive films based on highly aligned polyimide fibers containing hybrid materials of graphene nanoribbons and carbon nanotubes. Nanoscale **7**, 1037–1046 (2015)
79. Huang, L., Zhang, Z., Li, Z., Chen, B., Ma, X., Dong, L., Peng, L.-M.: Multifunctional graphene sensors for magnetic and hydrogen detection. ACS Appl. Mater. Interf. **7**, 9581–9588 (2015)
80. Zhang, L., Fan, W., Liu, T.: Flexible hierarchical membranes of WS_2 nanosheets grown on graphene-wrapped electrospun carbon nanofibers as advanced anodes for highly reversible lithium storage. Nanoscale **8**, 16387–16394 (2016)
81. Yang, S.-Y., Chang, K.-H., Lee, Y.-F., Ma, C.-C.M., Hu, C.-C.: Constructing a hierarchical graphene–carbon nanotube architecture for enhancing exposure of graphene and electrochemical activity of Pt nanoclusters. Electrochem. Commun. **12**, 1206–1209 (2010)
82. Wang, Y.-S., Li, S.-M., Hsiao, S.-T., Liao, W.-H., Chen, P.-H., Yang, S.-Y., Tien, H.-W., Ma, C.-C.M., Hu, C.-C.: Integration of tailored reduced graphene oxide nanosheets and electrospun polyamide-66 nanofabrics for a flexible supercapacitor with high-volume-and high-area-specific capacitance. Carbon **73**, 87–98 (2014)
83. Rao, C. N. R., Maitra, U., Matte, H. S. S. R.: Synthesis, characterization, and selected properties of graphene. Graphene. 1–47 (2012)
84. Chen, W., Chen, Y., Cheng, Y., Zhang, W., Shao, M., Shen, Y., Wu, P., Zheng, B., Li, S., Zhang, W., Wu, J.: Three-dimensional multilayered interconnected network of conjugated carbon nanofibers encapsulated silicon/graphene oxide for lithium storage. J. Inorg. Organomet. Polym. Mater. **30**, 801–807 (2020)
85. Shan, C., Wang, Y., Xie, S., Guan, H.-Y., Argueta, M., Yue, Y.: Free-standing nitrogen-doped graphene-carbon nanofiber composite mats: electrospinning synthesis and application as anode material for lithium-ion batteries. J. Chem. Technol. Biotechnol. **94**, 3793–3799 (2019)
86. Zhao, Q., Liu, J., Li, X., Xia, Z., Zhang, Q., Zhou, M., Tian, W., Wang, M., Hu, H., Li, Z., Wu, W., Ning, H., Wu, M.: Graphene oxide-induced synthesis of button-shaped amorphous Fe_2O_3/rGO/CNFs films as flexible anode for high-performance lithium-ion batteries. Chem. Eng. J. (Amsterdam, Neth.). **369**, 215–222 (2019)
87. Yang, C., Lan, J.-L., Ding, C., Wang, F., Siyal, S.H., Yu, Y., Yang, X.: Three-dimensional hierarchical ternary aerogels of ultrafine TiO_2 nanoparticles@porous carbon nanofibers-reduced graphene oxide for high-performance lithium-ion capacitors. Electrochim Acta. **296**, 790–798 (2019)
88. Wu, S., Han, Y., Wen, K., Wei, Z., Chen, D., Lv, W., Lei, T., Xiong, J., Gu, M., He, W.: Composite nanofibers through in-situ reduction with abundant active sites as flexible and stable anode for lithium ion batteries. Compos. B **161**, 369–375 (2019)
89. Xu, C., Liu, Z., Wei, T., Sheng, L., Zhang, L., Chen, L., Zhou, Q., Jiang, Z., Wang, L., Fan, Z.: Mesoporous single-crystalline MnOx nanofibers@graphene for ultra-high rate and long-life lithium-ion battery anodes. J. Mater. Chem. A. **6**, 24756–24766 (2018)

90. Wang, F., Cai, J., Yu, J., Li, C., Yang, Z.: Simultaneous electrospinning and electrospraying: fabrication of a carbon nanofibre/MnO/reduced graphene oxide thin film as a high-performance anode for Lithium-Ion batteries. ChemElectroChem **5**, 51–61 (2018)
91. Kumar, A., Prasada Rao, T., Jayakumar, O.D., Nazri, G.A., Naik, V.M., Naik, R.: Improved electrochemical performance of Li_2FeSiO_4/CNF/rGO nanocomposites for lithium ion batteries. Solid State Ionics **325**, 43–47 (2018)
92. Jamal, H., Kang, B.-S., Lee, H., Yu, J.-S., Lee, C.-S.: Comparative studies of electrochemical performance and characterization of TiO_2/graphene nanocomposites as anode materials for Li-secondary batteries. J. Ind. Eng. Chem. (Amsterdam, Neth.). **64**, 151–166 (2018)
93. Gao, Q., Yuan, Z., Dong, L., Wang, G., Yu, X.: Reduced graphene oxide wrapped $ZnMn_2O_4$/carbon nanofibers for long-life lithium-ion batteries. Electrochim. Acta. **270**, 417–425 (2018)
94. Gao, H., Sun, H., Zhao, A., Wang, L., Liu, N.: Co_3O_4 nanofibers modified by graphene as high-performance anode materials for lithium-ion batteries. Int. J. Electrochem. Sci. **13**, 8666–8672 (2018)
95. Wu, Y., Chen, Y., Lin, J., Chu, R., Zheng, J., Wu, C., Guo, H.: Three-dimensional tin dioxide-graphene composite nanofiber membrane as binder-free anode for high-performance lithium-ion batteries. J. Mater. Sci. **52**, 8097–8106 (2017)
96. Wu, C., Lin, J., Chu, R., Zheng, J., Chen, Y., Zhang, J., Guo, H.: Reduced graphene oxide as a dual-functional enhancer wrapped over silicon/porous carbon nanofibers for high-performance lithium-ion battery anodes. J. Mater. Sci. **52**, 7984–7996 (2017)
97. Wei, W., Jia, F., Wang, K., Luo, B., Qu, P., Xu, M.: Synthesis of graphene/$Ca_2Ge_7O_{16}$ nanofibers composite as anode materials for lithium-ion batteries. Mater. Lett. **196**, 157–160 (2017)
98. Wang, J., Li, C., Yang, Z., Chen, D.: Chemical vapor deposition-assisted fabrication of a graphene-wrapped MnO/carbon nanofibers membrane as a high-rate and long-life anode for lithium ion batteries. RSC Adv. **7**, 50973–50980 (2017)
99. Wang, J., Huang, T., Yin, W.: $Mn_{0.5}Co_{2.5}O_4$ nanofibers intercalated into graphene frameworks with mesoporous structure for batteries and supercapacitors. Energy Technol. (Weinheim, Ger.). **5**, 2055–2064 (2017)
100. Wang, F., Zhu, P., Pan, J., Li, C., Yang, Z.: A facile electrospinning and electrospraying synchronization technique for preparation of high performance MnO/C@rGO composite anodes for lithium storage. RSC Adv. **7**, 48294–48302 (2017)
101. Thirugunanam, L., Kaveri, S., Etacheri, V., Ramaprabhu, S., Dutta, M., Pol, V.G.: Electrospun nanoporous TiO_2 nanofibers wrapped with reduced graphene oxide for enhanced and rapid lithium-ion storage. Mater. Charact. **131**, 64–71 (2017)
102. Qiao, H., Xia, Z., Fei, Y., Cai, L., Cui, R., Cai, Y., Wei, Q., Yao, Q.: Electrospinning combined with hydrothermal synthesis and lithium storage properties of $ZnFe_2O_4$-graphene composite nanofibers. Ceram. Int. **43**, 2136–2142 (2017)
103. Peng, C.J., Wang, L., Li, Q.W., Li, Y.Y., Huo, K., Chu, P.K.: Ge@CNFs Anchored on 3D graphene foam for binder-free and high-efficiency anodes in Li-Ion batteries. ChemElectroChem **4**, 1002–1006 (2017)
104. Ma, X., Hou, G., Ai, Q., Zhang, L., Si, P., Feng, J., Ci, L.: A heart-coronary arteries structure of carbon nanofibers/graphene/silicon composite anode for high performance lithium ion batteries. Sci. Rep. **7**, 9642 (2017)
105. Li, D., Guo, E., Lu, Q., Ji, X., Wei, M.: Synthesis of one-dimensional graphene-encapsulated TiO_2 nanofibers with enhanced lithium storage capacity for lithium-ion batteries. J. Solid State Electrochem. **21**, 2313–2320 (2017)
106. Joshi, B., Lee, J.-G., Samuel, E., Jo, H.S., Kim, T.-G., Swihart, M.T., Yoon, W.Y., Yoon, S.S.: Supersonically blown reduced graphene oxide loaded Fe-Fe_3C nanofibers for lithium ion battery anodes. J. Alloys Compd. **726**, 114–120 (2017)
107. Hu, R., Zhang, H., Bu, Y., Zhang, H., Zhao, B., Yang, C.: Porous Co_3O_4 nanofibers surface-modified by reduced graphene oxide as a durable, high-rate anode for lithium ion battery. Electrochim. Acta. **228**, 241–250 (2017)

108. He, Z., Wu, X., Yi, Z., Wang, X., Xiang, Y.: Silicon/graphene/carbon hierarchical structure nanofibers for high performance lithium ion batteries. Mater. Lett. **200**, 128–131 (2017)
109. He, J., Zhao, S., Lian, Y., Zhou, M., Wang, L., Ding, B., Cui, S.: Graphene-doped carbon/Fe_3O_4 porous nanofibers with hierarchical band construction as high-performance anodes for lithium-ion batteries. Electrochim. Acta. **229**, 306–315 (2017)
110. Dufficy, M.K., Huang, S.-Y., Khan, S.A., Fedkiw, P.S.: Effects of composition and structure on the performance of tin/graphene-containing carbon nanofibers for Li-ion anodes. RSC Adv. **7**, 15428–15438 (2017)
111. Cho, S.-H., Jung, J.-W., Kim, C., Kim, I.-D.: Rational design of 1-D Co_3O_4 nanofibers@Low content graphene composite anode for high performance Li-Ion batteries. Sci. Rep. **7**, 45105 (2017)
112. Kim, Y.S., Shoorideh, G., Zhmayev, Y., Lee, J., Li, Z., Patel, B., Chakrapani, S., Park, J.H., Lee, S., Joo, Y.L.: The critical contribution of unzipped graphene nanoribbons to scalable silicon–carbon fiber anodes in rechargeable Li-ion batteries. Nano Energy **16**, 446–457 (2015)
113. Pham-Cong, D., Kim, J.Y., Park, J.S., Kim, J.H., Kim, J.-P., Jeong, E.-D., Kim, J., Jeong, S.-Y., Cho, C.-R.: Conductive framework supported high rate performance of SnO_2 hollow nanofibers for lithium battery anodes. Electrochim. Acta. **161**, 1–9 (2015)
114. Wu, H., Chan, G., Choi, J.W., Ryu, I., Yao, Y., McDowell, M.T., Lee, S.W., Jackson, A., Yang, Y., Hu, L., Cui, Y.: Stable cycling of double-walled silicon nanotube battery anodes through solid–electrolyte interphase control. Nat. Nanotechnol. **7**, 310–315 (2012)
115. Wu, P., Du, N., Zhang, H., Yu, J., Qi, Y., Yang, D.: Carbon-coated SnO_2 nanotubes: template-engaged synthesis and their application in lithium-ion batteries. Nanoscale **3**, 746–750 (2011)
116. Chen, T., Pan, L., Loh, T.A.J., Chua, D.H.C., Yao, Y., Chen, Q., Li, D., Qin, W., Sun, Z.: Porous nitrogen-doped carbon microspheres as anode materials for lithium ion batteries. Dalton Trans. **43**, 14931–14935 (2014)
117. Zheng, F., Yang, Y., Chen, Q.: High lithium anodic performance of highly nitrogen-doped porous carbon prepared from a metal-organic framework. Nat. Commun. **5**, 5261 (2014)
118. Zhang, H., Sun, X., Huang, X., Zhou, L.: Encapsulation of α-Fe_2O_3 nanoparticles in graphitic carbon microspheres as high-performance anode materials for lithium-ion batteries. Nanoscale **7**, 3270–3275 (2015)
119. Pham-Cong, D., Park, J.S., Kim, J.H., Kim, J., Braun, P.V., Choi, J.H., Kim, S.J., Jeong, S.Y., Cho, C.R.: Enhanced cycle stability of polypyrrole-derived nitrogen-doped carbon-coated tin oxide hollow nanofibers for lithium battery anodes. Carbon **111**, 28–37 (2017)
120. Zhang, X., Huang, X., Zhang, X., Xia, L., Zhong, B., Zhang, T., Wen, G.: Flexible carbonized cotton covered by graphene/Co-doped SnO_2 as free-standing and binder-free anode material for lithium-ions batteries. Electrochim. Acta. **222**, 518–527 (2016)
121. Chen, C., Xu, H., Zhou, T., Guo, Z., Chen, L., Yan, M., Mai, L., Hu, P., Cheng, S., Huang, Y., Xie, J.: Integrated intercalation-based and interfacial sodium storage in graphene-wrapped porous $Li_4Ti_5O_{12}$ nanofibers composite aerogel. Adv. Energy Mater. **6**, 1600322 (2016)
122. Ryu, W.-H., Yoon, T.-H., Song, S.H., Jeon, S., Park, Y.-J., Kim, I.-D.: Bifunctional composite catalysts using Co_3O_4 nanofibers immobilized on nonoxidized graphene nanoflakes for high-capacity and long-cycle Li–O_2 batteries. Nano Lett. **13**, 4190–4197 (2013)
123. Wang, Y., Ding, X., Wang, F., Li, J., Song, S., Zhang, H.: Nanoconfined nitrogen-doped carbon-coated MnO nanoparticles in graphene enabling high performance for lithium-ion batteries and oxygen reduction reaction. Chem. Sci. **7**, 4284–4290 (2016)
124. Gao, G., Shi, L., Lu, S., Gao, T., Li, Z., Gao, Y., Ding, S.: Ethylene glycol-mediated rapid synthesis of carbon-coated $ZnFe_2O_4$ nanoflakes with long-term and high-rate performance for lithium-ion batteries. Dalton Trans. **47**, 3521–3529 (2018)
125. Wang, S., Zhang, H., Zhang, D., Ma, Y., Bi, X., Yang, S.: Vertically oriented growth of MoO_3 nanosheets on graphene for superior lithium storage. J. Mater. Chem. A. **6**, 672–679 (2018)
126. Chao, D., Lai, C.-H., Liang, P., Wei, Q., Wang, Y.-S., Zhu, C., Deng, G., Doan-Nguyen, V.V.T., Lin, J., Mai, L., Fan, H.J., Dunn, B., Shen, Z.X.: Sodium vanadium fluorophosphates (NVOPF) array cathode designed for high-rate full sodium ion storage device. Adv. Energy Mater. **8**, 1800058 (2018)

127. Zhu, S., Li, L., Liu, J., Wang, H., Wang, T., Zhang, Y., Zhang, L., Ruoff, R.S., Dong, F.: Structural directed growth of ultrathin parallel birnessite on β-MnO_2 for high-performance asymmetric supercapacitors. ACS Nano **12**, 1033–1042 (2018)
128. Smith, P.F., Deibert, B.J., Kaushik, S., Gardner, G., Hwang, S., Wang, H., Al-Sharab, J.F., Garfunkel, E., Fabris, L., Li, J., Dismukes, G.C.: Coordination geometry and oxidation state requirements of corner-sharing MnO_6 octahedra for water oxidation catalysis: an investigation of manganite (γ-MnOOH). ACS Catal. **6**, 2089–2099 (2016)
129. Mitchell, J.F., Argyriou, D.N., Berger, A., Gray, K.E., Osborn, R., Welp, U.: Spin, charge, and lattice states in layered magnetoresistive oxides. J. Phys. Chem. B. **105**, 10731–10745 (2001)
130. Guo, J., Zhu, H., Sun, Y., Tang, L., Zhang, X.: Pie-like free-standing paper of graphene paper@Fe_3O_4 nanorod array@carbon as integrated anode for robust lithium storage. Chem. Eng. J. **309**, 272–277 (2017)
131. Zhu, X., Song, X., Ma, X., Ning, G.: Enhanced electrode performance of Fe_2O_3 nanoparticle-decorated nanomesh graphene as anodes for lithium-ion batteries. ACS Appl. Mater. Interf. **6**, 7189–7197 (2014)
132. Yin, L., Gao, Y.J., Jeon, I., Yang, H., Kim, J.-P., Jeong, S.Y., Cho, C.R.: Rice-panicle-like γ-Fe_2O_3@C nanofibers as high-rate anodes for superior lithium-ion batteries. Chem. Eng. J. **356**, 60–68 (2019)

Graphene Nanofiber-Based Composites for Fuel Cell Application

Benalia Kouini and Hossem Belhamdi

Abstract Graphene is becoming very important in recent years, due to its interesting electrical, thermal, mechanical, and electrochemical properties. With these advantages, graphene has been used as a substance to incorporate various kinds of functional materials. Among them, graphene-based polymer nanocomposites have attracted a great deal of interest due to their new properties or enhanced performance during the last few years. This is because graphene-based polymer nanocomposites are used for electronic, optical, and electrochemical applications owing to their advantages. This chapter review discusses the preparation methods of graphene-polymer nanocomposites, some examples of graphene/polymer composites, the electrochemical behavior of graphene, and summarizes the recent research and development on Graphene nanofiber-based composites fuel cells.

1 Introduction

Graphene, a form of carbon experimentally demonstrated in 2004, [1] consists of a monolayer of carbon atoms sp^2 bonded into a two-dimensional sheet. Graphene is a basic building block for graphitic materials of all other dimensionalities as illustrated in Fig. 1 Graphene possesses unique electronic properties, including ambipolar electric field effect, extremely high mobility of charge carriers (up to 105 $cm^2\ V^{-1}\ s^{-1}$), mass-less electron, quantum Hall effects (QHE) even at room temperature, and electron wave propagation within a one atom thick layer [2, 3]. The unique properties of graphene have sparked widespread interest and investigations

B. Kouini (✉)
Laboratory of Coatings, Materials and Environment, M'Hamed Bougara University, 35000 Boumerdes, Algeria
e-mail: kouinib@univ-boumerdes.dz

H. Belhamdi
Research Unit: Materials, Processes and Environment (RU/MPE), M'Hamed Bougara University, 35000 Boumerdes, Algeria
e-mail: h.belhamdi@univ-boumerdes.dz

S. K. Tiwari et al. (eds.), *Electrospinning of Graphene*, Carbon Nanostructures,
https://doi.org/10.1007/978-3-030-75456-3_6

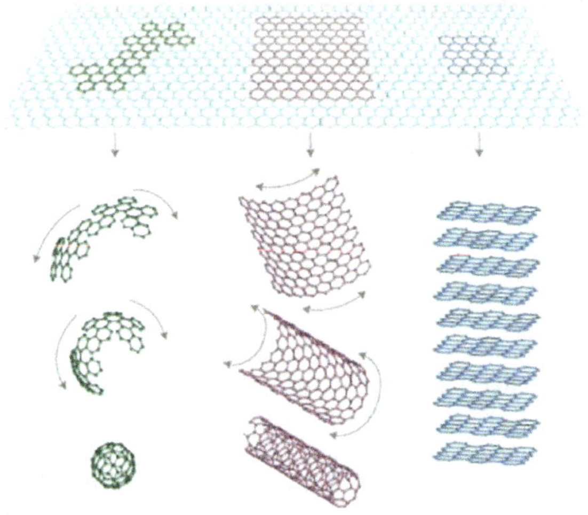

Fig. 1 Graphene is the building block of all graphitic forms: 0-dimensional buckyballs, 1-dimensional carbon nanotube, or 3-dimensional graphite (Reproduced with permission from ref [2]. Copyright 2007, Springer Nature)

in laboratories around the world. But despite the surge in interest, the chemical and physical understanding of graphene is still evolving as are the potential applications.

Graphene shows a very high surface area (B2600 m^2 g^{-1}), much larger than the surface areas of graphite (B10 m^2 g^{-1}) and carbon nanotubes (1300 m^2 g^{-1}) [4]. These remarkable properties make graphene promising in applications such as polymer-composite materials, [4, 5] paper-like materials, [6] photo-electronics, [7–9] field-effect transistors, [10] electromechanical systems, [11, 12] sensors and probes, [13–15] hydrogen storage, [16] and of course electrochemical energy systems [17].

Also, the discovery of graphene with its combination of extraordinary physical properties and the ability to be dispersed in various polymer matrices has created a new class of polymer nanocomposites [18].

One of the most common graphene-based nanocomposites is graphene-based polymer nanocomposites, which show superior thermal, electrical, mechanical, optical, and electrochemical properties compared with the neat polymer or graphene. Therefore, graphene-based polymer nanocomposites are intensively explored in applications such as supercapacitors, sensing platforms, fuel cells, solar cells, and so on [19].

2 Preparation Methods of the Graphene-Polymer Nanocomposites

The properties of graphene-based polymer nanocomposites are dependent upon the processing conditions in the fabrication of graphene/polymer nanocomposites. The functionality of graphene components is critical to lower filler loading rate, make them highly dispersed and organized sheets within a polymer matrix to enhance the properties of nanocomposites. In particular, the mechanical properties depend on the specific surface area, aspect ratio, organization, and loading content of graphene materials. The dispersion, interfacial strength, affinity of components, and spatial organization are all of great importance in determining the final stiffness, strength, toughness, and elongation of polymer nanocomposites under various loading conditions [20–24]. The pretreatment procedures and the fabrication methods control the fine morphology and physical/chemical properties of graphene-based polymer nanocomposites. For various graphene-polymer nanocomposites known to date, the extent of dispersion and exfoliation of graphitic layers is controlled by the applied shear force, temperature, and solvent polarity. Effective control of restacking, wrinkling, and aggregation of graphene sheets is required for the development of functional nanocomposites with high performance. Extremely flexible and high-aspect-ratio graphene components are prone to random wrinkling, buckling, or folding during processing, which dramatically affects the ultimate performance. In the case of the post-treatment, the degree of dispersion can be further influenced by the hydrophobic nature of reduced graphene oxide sheets and dewetting processes at the interfaces.

The choice of fabrication methods is determined by the surface functionalization of integrated graphitic sheets. Generally, traditional fabrication routines include solution-based processing [25–28] and melt-based processing [29–31]. Among the most popular approaches for chemical modification and assembly are in-situ polymerization, chemical grafting, latex emulsion blending, layer-by-layer (LbL) assembly, and directed assembly [25, 26, 32–34]. For the in-situ polymerization method, intercalated monomers within expanded graphite clusters can promote their efficient exfoliation into single sheets throughout the polymer matrix caused by catalysis reactions [27].

Solution processing maximizes filler dispersion in a polymer matrix by using pre-suspended single-layered graphene sheets. Different solvents (aqueous to organic) can be used to dissolve graphene materials, including graphene oxide and reduced graphene oxide materials. This approach has been widely exploited due to its high dispersion efficiency, facile and fast fabrication step, and a high level of control on component behavior. By the way, melt-based mixing is a solvent-free process in which applied mechanical shear force distributes the fillers in the polymer matrix using a screw extruder or a blending mixer [28, 35]. This method allows stacked graphite or reduced graphene oxide sheets to be exfoliated into a viscous polymer melt by suppressing unfavorable interactions and inducing component dispersion. Melt mixing is recognized as a practical approach that can be adapted to graphene-based polymer nanocomposites.

3 Examples of Graphene/Polymer Composites

3.1 Graphene/Epoxy Composites

Graphene has been incorporated into epoxy-based materials to enhance various functional properties. Graphene's excellent mechanical properties make it is a good candidate for reinforcement in nanocomposites. Bortz et al. [36] investigated the effect of graphene concentration on the mechanical properties of graphene/epoxy nanocomposites. The tensile modulus of unmodified epoxy increased by ≈12% at only 0.1 wt.% loading of graphene oxide. Wajid et al. [37] used solution mixing to fabricate graphene/epoxy composites. The strength and modulus increased by 38 and 37% respectively at only 0.46 vol% graphene loading. By using a simple solution mixing technique, Galpaya et al. [38] fabricated a graphene oxide/epoxy nanocomposite with enhanced mechanical properties. At only 0.1 wt.% additionof graphene oxide, fracture toughness increased by ≈50%. The graphene oxide sheets in the composite disturbed and deflected the crack propagation. The elastic modulus increased by ≈35% from the neat epoxy with only 0.5 wt.% loading of graphene oxide. Shen et al. [39] investigated the tribological properties of epoxy nanocomposites reinforced with graphene oxide at low loading (0.05–0.5 wt.%). The wear resistance significantly increased with the addition of graphene oxide, such that, at 0.5 wt.% loading, the specific wear rate was reduced by 90.0–94.1% compared to the neat polymer.

Electrical conductivity in nanocomposites is achieved by the formation of a continuous network of the conductive fillers [40, 41]. Therefore, because graphene has a very high-aspect-ratio, the percolation threshold is achieved at very low graphene loading [42]. Wajid et al. [37] used sonication and shear mixing methods to disperse graphene into the epoxy matrix and achieved the percolation threshold at 0.088 vol%. In another study, the composites were fabricated by the incorporation of functionalized graphene into the epoxy matrix by an in-situ process. The percolation threshold was reached only after 0.52 wt.% loading of graphene, and electrical conductivity increased from 1^{-10} to 1^{-2} S cm^{-1} by ≈9 vol% loading of reduced graphene oxide. Zhao et al. [43] conducted a thorough investigation into the properties of epoxy composite filled with epoxide-functionalized graphene (G-EP). The tensile strength and Young's modulus increased by 116% and 96%, respectively, compared to the polymer at only 1 wt.% loading of G-EP. The percolation threshold was reached at only 0.33 wt.% loading, and electrical conductivity increased from about 1^{-17} to 1^{-2} $S^{cm^{-1}}$ at ≈2 vol% loading. Thermal conductivity also improved by 189% at 10 wt.% G-EP loading. For a more comprehensive review of graphene/epoxy nanocomposites, the reader is referred to a review by Rasheed et al. [44].

3.2 Graphene/Cellulose Composites

Graphene has been incorporated in cellulose composites to enhance mechanical and electrical properties. GO and cellulose was dissolved in N-methyl morpholine-N-oxide (NMMO) monohydrate to prepare the composite films [45]. Despite being brittle, the produced graphene/cellulose composites showed improved thermal and electrical properties. Nguyen et al. [46] used reduced graphene oxide sheets (RGO) and amine-modified nano fibrillated cellulose (A-NFC) to produce composites with enhanced properties. These composites showed good electrical conductivity of up to 71.8 Sm^{-1} and tensile strength of up to 273 MPa. Graphenecellulose paper (GCP) membranes are used to fabricate flexible supercapacitors [47]. These composites displayed high electrical conductivity stability with a decrease of only 6% after being bent 1000 times. The capacitance of these supercapacitors per geometric area is 81 mF cm^{-2}, which is equivalent to a gravimetric capacitance of 120 F g^{-1} of graphene, and these supercapacitors did not lose any capacitance after 5000 cycles. Manman et al. [48] reported cellulose/graphene composite hydrogels prepared from ionic liquids (IL). They used Vitamin C to prepare RGO directly in IL. These hydrogels showed enhanced mechanical and thermal properties. At only 0.5 wt.% doping of RGO in cellulose composite, Young's modulus was improved more than four times.

3.3 PVA/Graphene Nanocomposites

Zhao et al. [49] reported the preparation of graphene nanosheets and poly (vinyl alcohol) (PVOH) via a facial aqueous solution. In their study, they showed that Graphene, flat carbon nanosheets, has generated huge activity in many areas of science and engineering due to its unprecedented physical and chemical properties. With the development of wide-scale applicability including facile synthesis and high yield, this exciting material is ready for its practical application in the preparation of polymer nanocomposites. Here they have reported that nanocomposites based on fully exfoliated graphene nanosheets and poly (vinyl alcohol) (PVA) are prepared via a facial aqueous solution. Significant enhancement of mechanical properties of the graphene/PVA composites is obtained at low graphene loading; that is, a 150% improvement of tensile strength and a nearly 10 times increase of Young's modulus are achieved at a graphene loading of 1.8 vol %. The comparison between the experimental results and theoretical simulation for Young's modulus indicates that the graphene nanosheets in the polymer matrix are mostly dispersed randomly in the nanocomposite films. Liang et al. [50] also studied poly (vinyl alcohol) (PVOH)/-graphene nanocomposites. In this study, they used water as a solvent to fuse GO into the PVOH matrix. At only 0.7 wt.% GO loading, tensile strength increased by 76%, and Young's modulus by 62%.

3.4 Polyurethane (PU)/Graphene Composites

Lee et al. [51] reported the study of waterborne polyurethane (WPU)/graphene sheets (FGS) nanocomposites prepared using an in-situ method. The electrical conductivity of the nanocomposites was increased 10^5-fold compared with that of pure WPU. This was attributed to the high dispersity and homogeneity of graphene sheets in the WPU matrix. They obtained a percolation threshold of graphene sheets at only 2 wt.%. Liang et al. [52] used thermoplastic polyurethane (TPU) to fabricate nanocomposites with three differently processed graphene samples. They were able to show that the rate of thermal degradation for thermoplastic polyurethane/isocyanate modified graphene composites are much higher than that of the sulfonated graphene and reduced graphene-based TPU nanocomposites. The TPU/-graphene nanocomposites doped with only 1 wt.% sulfonated graphene showed enhanced infrared-triggered actuation performance. The tests showed that the composite could contract and lift a 21.6 g weight up to 3.1 cm with 0.21 N of force when exposed to IR. The mechanical properties of this composite were also increased. The tensile strength of TPU/sulfonated graphene nanocomposites with 1 wt.% graphene was increased by 75% at a strain of 100%, and Young's modulus was enhanced by 120% [52]. However, IR-triggered actuation performance of isocyanate modified graphene/TPU nanocomposites showed inferior results.

3.5 Graphene/Polyethylene Terephthalate (PET) Nanocomposites

The melt intercalation technique was used to prepare PET/-graphene composites by Zhang et al. [53] Transmission electron microscopy analysis showed that graphene nanosheets were evenly distributed in the PET matrix. This criterion was confirmed by the high electrical conductivity values they got from the PET/-graphene composites. The threshold percolation was only 0.47 vol% loading and with graphite 2.4 vol%. The high electrical conductivity of 2.11 Sm^{-1} was achieved with the addition of only 3.0 vol% graphene.

3.6 Polycarbonate (PC)/Graphene Nanocomposites

Kim et al. [54] studied PC composites doped with graphite and functionalized graphene sheets (FGS) by the melt intercalation technique. In this study, they used melt rheology to study the viscoelastic properties of the composites. They annealed the composites for 10,000 s and observed that the composites manifest a solid-like state above the percolation threshold, which was around 1 wt.% for the FGS, while around 3 wt.% for graphite loading. In terms of electrical properties, FGS/PC showed a lower percolation threshold as compared to graphite/PC composites. Both graphite and FGS fillers lead to improved PC composites stiffness and dimensional stability. The composites also exhibited good barrier properties. The permeability of nitrogen and helium of the PC composites is significantly reduced by the incorporation of both FGS and graphite. Moreover, FGS showed potential for application in gas separation processes as it was more effective against molecules with large kinetic diameters.

3.7 Polystyrene (PS)/Graphene Nanocomposites

Polystyrene/graphene composites have been extensively studied for various applications. Stankovich et al. [55] reported PS/isocyanate modified graphene composites prepared using a solution blending method in DMF. They obtained a percolation threshold for the electrical conductivity at 0.1 vol% GO in PS. The low percolation value is due to the homogeneous dispersion and an extremely large aspect ratio of graphene sheets. The PS/graphene composite showed an electrical conductivity of $\approx$0.1 to $\approx$1 Sm^{-1} at 2.5 vol% loading. Liu et al. [56] prepared a PS/GNP composite in ionic liquids. The electrical conductivity increased from 10^{-14} Sm^{-1} for pure PU to 5.77 Sm^{-1} with the addition of 0.38 vol% GNP, which is 3–15 times higher than that of polystyrene composites filled with single-walled carbon nanotubes. They also studied the thermal stability of PS/GNP composite and pure polystyrene. The degradation temperature of the PS/GNP composite was about 50 °C higher than that of pure PS. This enhancement is due to the strong interaction of GNP and the polymer matrix at the interface, which leads to a decrease of polymer chain mobility near the interface and, hence, the increase in thermal stability.

3.8 Other Graphene-Based Polymer Nanocomposites

Poly (ε-caprolactone) (PCL)/graphene oxide composite was prepared using in-situ polymerization [57]. The resulting nanocomposite shows excellent mechanical properties and robustness under bending. Poly (lactic acid) (PLA)/graphene nanocomposites were prepared using the response surface method [58], which showed that graphene loading had a significant effect on tensile strength. Liu et al. [59] fabricated graphene oxide (GO) reinforced epoxy resin nanocomposite by transferring GO from water to acetone. Incorporation of _1 wt.% of graphene oxide showed a significant improvement in flexural strength, flexural modulus, impact strength, and storage modulus. Mohamadi et al. prepared polymethyl methacrylate (PMMA)/graphene nanocomposite by sing in-situ polymerization [60]. Liang et al. fabricated polydiacetylene (PDA)/graphene nanocomposite by solution processing method [61]. The resulting nanocomposite showed excellent actuation character with controllable motion, fast response rate, and high-frequency resonance. Polyphenylene sulfide

(PPS)/graphene composite was prepared by the spraying method [62]. The resulting nanocomposite had seven times higher wear life than that of neat PPS.

Pang and his co-workers reported a novel conductive ultrahigh-molecular-weight polyethylene (UHMWPE) composite with a segregated and double percolated structure containing high-density polyethylene (HDPE) as carrier polymer for graphene nanosheets (GNS) [63].

4 Graphene-Based Electrochemical Energy Devices

Junbo Hou et al has well developed Graphene-based electrochemical energy devices [64]. The electrochemistry of graphene has been systematically investigated concerning surface chemistry and structure, heterogeneous charge transfer rate constants on redox species, electrochemical operating window, and electrocatalysis of molecules [65, 66]. The most notable feature of graphene is the very large electrochemically active surface area. Cyclic voltammetry shows 13 times higher capacitance on the graphene electrode than that on the bare glassy carbon (GC) [65]. Calculated from chronocoulometric curves at different electrodes for the reduction of 1 mM $K_3Fe(CN)_6$ with 2 M KCl, the area for different electrodes is graphene/GC (0.092 cm^2) > GC electrodes (0.0706 cm^2) > graphite/GC (0.0560 cm^2) [66]. Besides the very large physical surface area, the sp2-hybridized structure may contribute to the high capacitance value particularly for graphene. Finally, as with other carbon allotropes, [67], significant edge-plane-like defective sites are existing on the surface of graphene.

Several systems including $Ru(NH_3)_6^{3+/2+}$, $Fe(CN)_6^{3-/4-}$ $Fe^{3+/2+}$, and dopamine (DA) have been used for testing the electrochemistry of graphene because of their well-known sensitivity and selectivity to the electronic properties, surface microstructure, and surface chemistry of carbon electrodes [65, 67]. Voltammograms of these redox systems demonstrate fast charge transfer kinetics on a graphene electrode. Electrochemical impedance spectroscopy (EIS) shows that the graphene electrode possesses smaller charge transfer resistance compared with the graphite/GC and GC electrodes. In addition to its favorable surface characteristics, the electrochemical operating window for a graphene electrode in 0.1 M pH 7.0 PBS is 2.5 V, which is comparable to that for graphite and GC electrodes [66]. Finally, graphene prepared from graphite generally does not contain any heterogeneous impurities which are a significant advantage over CNTs [68]. The following sections discuss the implications of these and other graphene characteristics for electrochemical device applications including fuel cells, supercapacitors, and batteries.

4.1 Fuel Cells

Polymer electrolyte membrane (PEM) fuel cells are electrochemical energy conversion devices, which have attracted considerable interest as power sources for mobile and stationary applications, with active research and development programs all over the world focused on this field in recent decades [69, 70]. However, despite this effort, the cost and durability issues related to the materials and components still hinder the commercialization of PEM fuel cells. Graphene sheets, with their unique nanostructure and properties, are promising alternatives for several of the critical PEM fuel cell materials. As a carbon allotrope, the most obvious use for graphene is as catalyst support. Generally, the catalyst support materials should have particular properties such as (i) high specific surface area, improving the dispersion of catalytic metals, (ii) chemical stability under oxidative and reductive conditions at relevant temperatures (150 °C or less for PEM fuel cells), (iii) high electrochemical stability under fuel cell operating conditions, (iv) high conductivity, and (v) easy recovery of Pt in the used catalyst [71].

For porous electrodes, such as those employed in PEM fuel cells, the specific surface area of the catalyst support is a critical factor. For graphene, the theoretical specific surface area of graphene is about 2600 $m^2\ g^{-1}$, but the specific area of multilayer graphene decreases. The limiting case is graphite, having a specific surface area of only 10 $m^2\ g^{-1}$. If the graphene-supported catalysts are prepared in one step, the chemical reductions of graphene oxide and the metal salts occur simultaneously. With this process, there are most likely single-, double-, and few-layer graphene sheets as well as thin graphite films co-existing in the resulting supported catalyst. Thus, the actual specific surface area of the supported catalysts will be far below the theoretical value. Even so, the produced catalysts exhibit high electrocatalytic activity [72–74].

In addition to the active surface area, the durability of the catalyst and the support is an important issue in fuel cell applications. There are two ways by which the catalyst surface area is lost. One is coalescence sintering in which two catalyst clusters touch, and then fuse into one larger cluster.

The other is Ostwald ripening in which atoms evaporated from one cluster transfer to another, creating one larger cluster [71]. Coalescence sintering is enhanced by carbon corrosion which occurs under a high potential and acid environment (i.e., the PEM fuel cell condition). In this process, the carbon support is oxidized into CO_2 or CO causing the catalyst support to collapse and thus causing coalescence sintering. Also, when the carbon support is oxidized into surface oxygen-containing species, the catalyst–carbon interactions will be weakened also facilitating coalescence sintering. Graphene may improve catalyst durability by providing a more stable support material and by strengthening the interaction of the catalyst with the support.

Finally, a high-performance PEM fuel cell electrode requires the formation of numerous triple-phase boundaries (TPBs) to create efficient reaction sites at nanoscale or microscale. The TPB is accessible for mass transfer (reactants and products), protons, and electrons, and provides a site where electrochemical reactions can occur. The microstructure of the typical catalyst layer in the PEM fuel cells can be depicted as catalyzed carbon particles flooded with the electrolyte to form agglomerates covered with a thin film of electrolyte. The reactant gas first passes through the channels between the agglomerates, diffuses through the ionomer thin film and thereafter in the agglomerates, and then reaches the TPBs [75].

There are two main pore structures: the primary pores in the agglomerates with characteristic lengths on the order of nanometers or tens of nanometers; and the secondary pores between agglomerates with characteristic lengths from nanometer to a micrometer. These pore structures are essential for gas and water transport. The unique planar geometry of graphene coupled with its high electrical conductivity provides an opportunity to create particularly promising TPB architectures.

Research efforts to take advantage of the unique features of graphene to produce supported catalysts with enhanced activity, increased durability, and high-performance electrode architectures are discussed in the following sections.

4.1.1 Enhanced Electrocatalytic Activity

The prospect for enhanced fuel cell catalysts using a high specific surface area graphene support was explored by preparing graphene–metal particle nanocomposites [72]. This work demonstrated that graphene oxide can be reduced by ethylene glycol (EG), and thus, graphene–metal particle nanocomposites can be prepared in one step in the water-EG system using graphene oxide as a precursor with metal nanoparticles (Au, Pt, and Pd). This approach has the additional advantage that the metal nanoparticles can be adsorbed on graphene oxide sheets and thus facilitates the catalytic reduction of graphene oxide with EG. The cyclic voltammogram of methanol oxidation on the prepared graphene–Pt nanocomposites demonstrate the potential application in direct methanol fuel cells. In subsequent work, using the reducing agent $NaBH_4$, composites of graphene nanosheets decorated by Pt nanoclusters were prepared via reduction of graphite oxide and H_2PtCl_6 in one pot [73]. The electrochemically active surface area (ECSA) of the graphene-based catalyst is 44.6 $m^2\ g^{-1}$ while that of Pt/Vulcan is 30.1 $m^2\ g^{-1}$, and further, the graphene-based catalyst shows superior catalytic performance toward methanol oxidation (Fig. 2) [73]. An alternative synthesis approach, the rapid microwave method, was also used to produce a reduced graphene oxide supported Pt [74]. It is also found that the increased hydroxyl species on the catalysts makes it CO resistant. Besides the direct methanol or ethanol fuel cells, graphene-supported platinum also shows the higher electrocatalytic activity toward oxygen reduction [76].

In addition to work with single-metal catalysts, bimetallic Pt–Ru nanoparticles have been synthesized on graphene sheets using the EG reduction method [77] an6.2d impregnation method [78].

Results showed that graphene-supported Pt and Pt–Ru nanoparticles demonstrate enhanced performance over the widely used Vulcan XC-72R carbon black supported

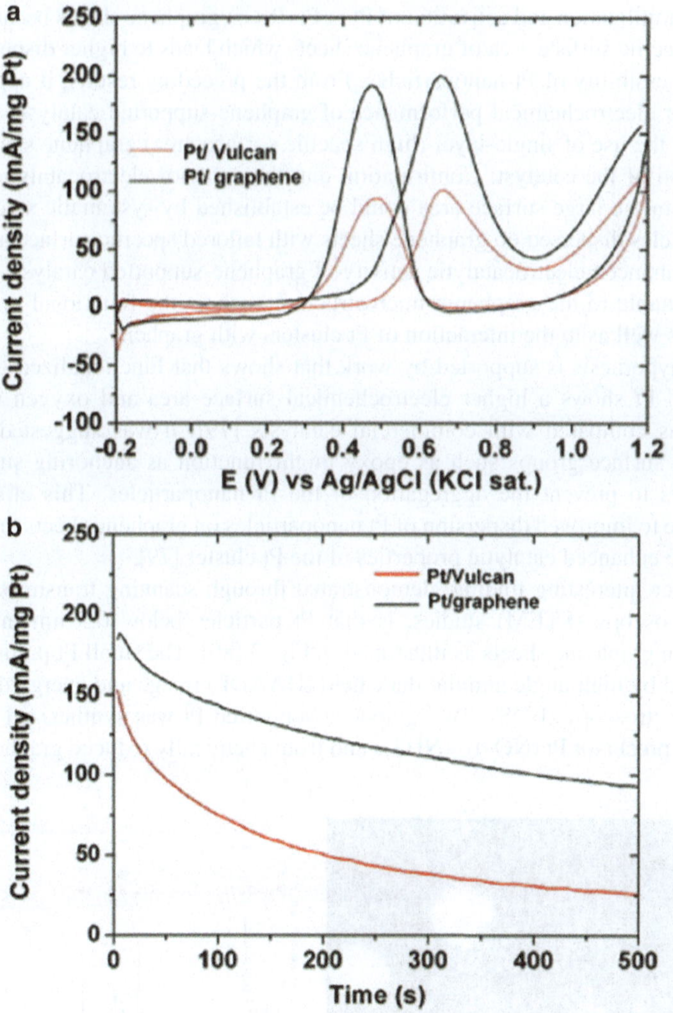

Fig. 2 Enhanced electrochemical activity toward methanol: **a** CV for Pt/graphene and Pt/Vulcan in nitrogen saturated aqueous solution of 0.5 M H_2SO_4 containing 0.5 M CH_3OH at a scan rate of 50 mV s^{-1}; **b** Chronoamperometric curves for the catalysts in the same solution at 0.6 V versus Ag/AgCl. (Reproduced with permission from ref [73]. Copyright © 2009 Elsevier B.V.)

catalyst for both methanol and ethanol electro-oxidation with regard to diffusion efficiency, oxidation potential, forward oxidation peak current density, and the ratio of the forward peak current density to the reverse peak current density [77]. Using the metal precursors (PtCl2 and RuCl3) and the solvent tetrahydrofuran(THF), a Pt–Ru catalyst with a high metal content of 80 wt.% was synthesized on the graphene sheets and showed superior electrochemical activity toward methanol oxidation compared with Pt–Ru/Vulcan XC-72R [78]. The enhanced catalytic activity was attributed to the higher utilization and activation of Pt or Pt–Ru on graphene sheets because of the higher specific surface area of graphene sheets which leads to higher dispersion and better accessibility of Pt nanoparticles. From the preceding results, it appears that the higher electrochemical performance of graphene-supported catalysts is associated with the use of single-layer (high specific surface area) graphene sheets in the production of the catalyst. Confirmation that the superior electrocatalytic activity arises from the large surface area could be established by systematic studies of Pt nanoparticles dispersed on graphene sheets with tailored specific surface areas.

The enhanced electrocatalytic activity of graphene-supported catalysts may also be attributable to the grapheme microstructure such as the functional groups and defects as well as to the interaction of Pt clusters with graphene.

This hypothesis is supported by work that shows that functionalized graphene-supported Pt shows a higher electrochemical surface area and oxygen reduction activity as compared with commercial catalysts [79]. It was suggested that the graphene surface groups such as epoxy might function as anchoring sites for Pt precursors to prevent the aggregation of the Pt nanoparticles. This effect could contribute to improved dispersion of Pt nanoparticles on graphene sheets thus giving rise to the enhanced catalytic properties of the Pt cluster [79].

Another interesting finding, demonstrated through scanning transmission electron microscopy (STEM) studies, is that Pt particles below 0.5 nm in size are formed on grapheme sheets as illustrated in Fig. 3 [80]. The small Pt particles were confirmed by high-angle annular dark-field (HAADF) image and energy dispersive X-ray spectroscopy (EDS). This graphene-supported Pt was synthesized from the platinum precursor Pt $(NO_2)_2\_(NH3)_2$ and from chemically reduced graphene using

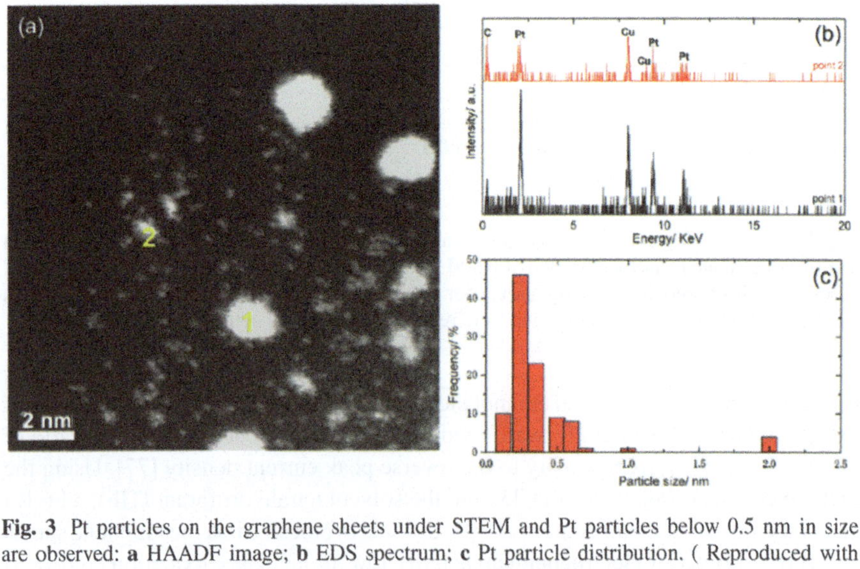

Fig. 3 Pt particles on the graphene sheets under STEM and Pt particles below 0.5 nm in size are observed: **a** HAADF image; **b** EDS spectrum; **c** Pt particle distribution. (Reproduced with permission from ref [80]. Copyright © 2009 American Chemical Society)

the impregnation method (Ar/H_2 at 400 °C). The resulting graphene-supported catalyst reveals an unusually high activity for methanol oxidation reaction compared to Pt/carbon black catalyst [80].

The author ascribed this enhanced electrocatalytic activity to the formation of dramatically smaller Pt particles. It is speculated that the strong interaction between Pt and graphene is responsible for the small Pt particles and thus might cause modulation in the electronic structure of the Pt clusters.

Unlike carbon supports (e.g., activated carbons), grapheme cannot exhibit micropores and deep cracks were deposited Pt nanoparticles would be isolated from the ionomer phase and thus unable to form electrochemically active TPBs. But carbon vacancies may exist in the produced graphene, especially in the chemically reduced graphene, where more charge, dangling bond, and functional groups can be entrapped. These features could lead to different localized electronic structures in the graphene sheets and interactions with catalyst precursors or metal clusters and therefore may impact the electrocatalytic activity. Interestingly, the relation of the peak current density and the scan rate in the cyclic voltammograms indicates that methanol diffuses faster on the surfaces of graphene sheets than that on the carbon black and graphite substrate [77].

Some of the preceding experimental studies are supported by theoretical studies showing that Pt bonds more strongly to functionalized graphene surfaces. From density functional theory (DFT) calculations, the gap between the highest occupied and lowest unoccupied molecular orbitals (the HOMO–LUMO gap) of hydrogen-terminated graphene is smaller than the zero-gap in infinite-size graphene. The small HOMO–LUMO gap is still retained after a Pt cluster binds to an edge of an sp^2 C surface to form two Pt–C bonds [81]. This theoretical calculation demonstrates that Pt clusters bond more strongly to the surface of the hydrogen-terminated graphene than to infinite-size graphene and that Pt clusters prefer the edge over the surface in the hydrogen-terminated graphene. In addition, oxygen-containing functional groups will not only function as anchoring sites for Pt precursors but also influence the position of the Pt cluster on the graphene sheets and the bonding energy with the sp^2 carbon in the graphene.

Further research is needed to understand the mechanism by which electrocatalytic activity is enhanced on graphene support. Specifically, studies comparing the performance of catalysts supported on defect-free and defect-rich graphene could help to explain the role of defects in the electrocatalytic activity. Likewise, systematic control of surface functional groups or of electronic structure (using nitrogen or boron doping) could yield a deeper understanding of the relative influence of these features and suggest areas for further improvement of graphene-supported catalysts.

4.1.2 Enhanced Durability

The unique structure of graphene may improve electrode durability by strengthening the interaction between the catalyst particles and the grapheme support. Using the accelerated degradation test (ADT) methods (potential step 1.4 V for 10 s to 0.85 V for 5 s), a poly (diallyldimethylammonium chloride) (PDDA) modified graphene-supported Pt was compared to Pt/CNT and E-TEK Pt/C [82]. After 46 h of testing, Pt/CNT and E-TEK Pt/C degraded by 75% based on both ECSA and oxygen reduction reaction (ORR), but graphene-supported Pt only degraded only 40% (ECSA) and 20% (ORR). In this test, the Ptmorphology did not change, and the improved durability can be attributed to more stable graphene support. In another durability test, Pt morphology did change and the influence of the support on the morphological stability was investigated [79]. After 5000 cycles of 0.6 to 1.1 V ADT, the average Pt particle size in a graphene-supported catalyst increased from 2 to 5.5 nm and more than 75% of the Pt particles remained smaller than 6.9 nm. For E-TEK carbon-supported catalysts, the average particle increased from 2.8 to 6.9 nm, and more than 45% of the particles were over 6.9 nm (Fig. 4). This result indicates that the interaction of Pt with graphene is stronger than the interaction with the E-TEK carbon support, thus leading to less Pt sintering on grapheme than on E-TEK carbon. The different observations between this study and the former study regarding Pt morphology changes may be attributable to the use of the reduction preparation method in the former study and of the impregnation heat treatment method in the latter study. In other words, the use of H_2 and high temperature also changes the structure

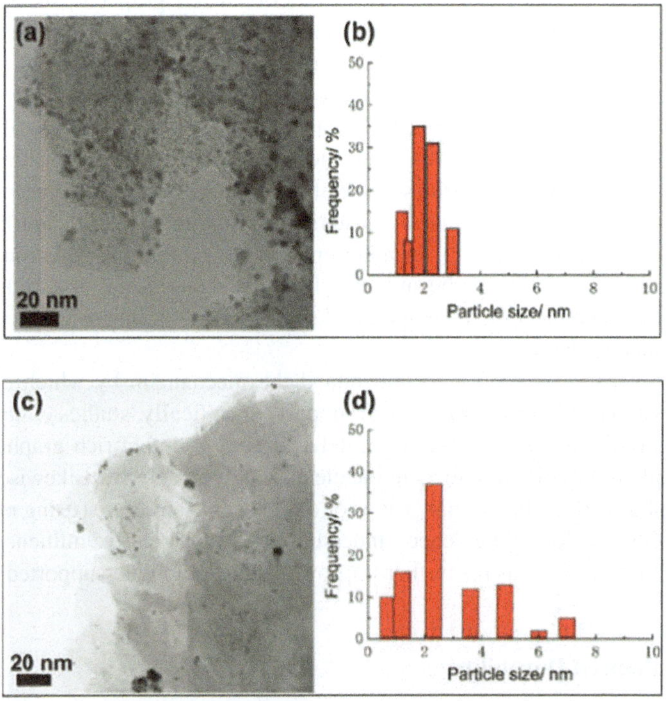

Fig. 4 TEM images of **a** E-TEK and **c** Pt-Graphene after 5000 CV degradation, and Pt nanoparticle size distribution diagrams on **b** E-TEK and **d** Pt-Graphene after 5000 CV degradation. (Reproduced with permission from ref [80]. Copyright © 2009 American Chemical Society)

of graphene sheets, creating H terminated graphene, and thus enhancing the Pt-sp^2 carbon bonding. As calculated by DFT, H terminated graphene has the advantage of enhancing the interactions between a Pt6 cluster and asp2 carbon surface [81].

As noted previously, nitrogen and boron can also be used to modify the electronic structure of the carbon support materials, thus improving the electrocatalytic activity and catalyst durability. The binding energy between a single platinum atom and several nitrogen-doped carbon grapheme structures was evaluated using DFT [83]. The nitrogen doping can double the binding energy, with the increase in binding energy proportional to the number and proximity of nitrogen atoms to the carbon-platinum bond. It should be noted that the addition of N atoms increases the number of dislocations in the lattice because N is prone to form pentagonal defects in the graphene structure [83].

4.1.3 Graphene-Based Electrodes

The use of 2-D grapheme as the catalyst support material in PEM fuel cells also creates opportunities to improve the structure of the catalyst layer although doing so requires consideration of the full range of catalyst layer properties-ionic transport, charge transport, and reactant transport in addition to ECSA. For example, one study has demonstrated an increase of nearly 80% in ECSA by further heat treatment of the catalyst (300 1C, 8 h).However, the fuel cell performance is lower than the untreated done, in which graphene-supported Pt-based fuel cell shows a maximum power of 161 mW cm^{-2} [76]. This can be attributed to the loss of proton conductivity of the Nafion ionomer in the catalyst layer during higher temperature treatment. Another key consideration in the development of graphene-based electrodes is the maintenance of good electronic conductance between graphene layers. As discussed above, nanostructured Pt particles dispersed on functionalized graphene can serve as spacers to prevent the re-agglomeration of graphene sheets. In addition, CNTs can be anchored on the graphene sheet to prevent face-to-face graphene agglomeration while simultaneously increasing the through-plane conductance. Polarization curves for the ORR in a PEM fuel cell electrode using a mix of graphene and CNT supported Pt as electrocatalysts exhibited performance of 540 mW cm-2 for the optimal case of equal fractions of each support (i.e., Pt/(50 wt.% graphenes + 50 wt.% CNT) [84]. This electrode structure is a combination of functionalized hybrid nanomaterials including three-, two-, and one-dimensional materials. While presenting some advantages, it is likely that the Pt/graphene sheets in this particular architecture pack together to form interstitial pores between the graphene sheets that are relatively inaccessible to electrolyte thus limiting the formation of TPBs. Recently, 3-D Pt-on-Pd bimetallic nanodendrites supported on graphene nanosheets have been constructed, representing a new type of graphene/metalhetero structure [85]. This structure, which exhibits a high electrochemically active area, consists of small single-crystal Pt nanoparticles supported on a Pd/graphene nanosheet with porous structure and good dispersion. Electrochemical tests show that the graphene/bimetallic nanodendrites hybrids demonstrate much higher electrocatalytic activity toward methanol oxidation than

Fig. 5 Strategy for novel electrode structure: bimetallic nanodendrite hybrid catalyzed graphene sheets can be covered by a thin proton ionomer film and assembled on the proton electrolyte membrane. (Reproduced with permission from ref [86]. Copyright© the Owner Societies 2011)

either platinum black (PB) or commercial E-TEK Pt/C catalysts [85]. Furthermore, as illustrated in Fig. 5, the bimetallic nanodendrites can be applied to construct a novel electrode structure based on graphene nano sheets, which could fully realize the benefit of 2D graphene while avoiding its disadvantage. The bimetallic nanodendrite hybrid catalyst can be deposited on the graphene sheets. Then, the catalyzed graphene sheets can be assembled onto the PEM in well-ordered macroscopic arrays using directional flow techniques [84]. A very thin ionomer film can then be sprayed onto the arrays to form the ionic pathway. This ideal structure provides for electron transport along with the graphene sheet and ion transport in the parallel electrolytic film connecting the membrane to reaction sites. In this structure, control of the distance between the graphene sheets is critical to gas and water transport. Compared to the traditional electrode structure, TPBs are much more accessible as there is no agglomerate structure, and catalyst utilization is improved.

4.1.4 Emerging Applications for Graphene in Fuel Cells

Graphene doped by nitrogen can function as a metal-free catalyst for the oxygen reduction reaction (ORR). In a recent study, nitrogen-doped graphene (N-graphene) was synthesized by chemical vapor deposition of methane in the presence of ammonia. As illustrated in Fig. 5.9, the resulting N-graphene shows a much better electrocatalytic activity, long-term operational stability, and tolerance to CO than platinum for the ORR [87]. According to an earlier report on CNTs doped by nitrogen, [88], carbon atoms adjacent to nitrogen dopants possess a substantially high positive charge density to counterbalance the strong electronic affinity of the nitrogen atom. A redox cycling process reduces the carbon atoms that naturally exist in an oxidized form, followed by reoxidation of the reduced carbon atoms to their preferred oxidized state upon O_2 absorption [88]. This is the ORR mechanism on the N-doped carbon electrodes. At the same time, the N-induced charge delocalization can change the chemisorption mode of O_2 and weaken the O–O bonding to facilitate ORR at the electrodes. Other work has also demonstrated, enhanced electrochemical activity and durability toward ORR on nitrogen-doped graphene [89]. The development of efficient, low cost, and stable N-doped graphene or CNT electrodes capable of replacing the expensive platinum-based electrocatalysts for ORR would be a revolutionary advance in PEM fuel cell technology.

Graphene sheets can also be used as composite fillers and conductive coating in fuel cell applications. PEM fuel cell performance can be improved by reducing ohmic losses in the membrane by either increasing ionic conductivity or decreasing the thickness of the membrane. Decreasing membrane thickness also facilitates water back diffusion allowing the anode to be operated without external humidification. But, the strength of the membranes becomes lower with the decrease in thickness. As a result, very thin membranes can bere inforced using polytetrafluoroethylene (PTFE) or CNT fillers. Graphene has a very high Young's modulus (B1.0 TPa) and could be used to reinforce the PEMs to dramatically increase their mechanical strength provided electrical shorting through the membrane can be avoided. Another potential application for graphene is as a coating on the bipolar plates to decrease the contact resistance and protect the plate material. Graphene is stable in the fuel cell environment and exhibits high thermal conductivity and electronic conductivity, and thus is potentially suitable for this application.

5 Electrospinning of Graphene

Electrospinning has emerged over the last decade as a promising nanotechnology-based approach to the design and development of intelligent and ultra-sensitive sensing systems [90]. The electrospinning technique is an effective method to produce nanofibres, nanobelts, Janus nanofibers, Janus nanobelts, hollow nanofibers, coaxial nanofibers, and coaxial nanoribbons with diameters ranging from micrometers to several nanometers using a wide variety of materials and improve the uniform dispersion of fillers and can also realize the effective alignment of fillers along with the oriented polymeric fibers [91, 92]. The electrospinning process consists of a high voltage power supply, a syringe pump, a metallic spinneret, and a conductive collector, as shown in Fig. 6 [90].

Graphene is usually hard to disperse uniformly in a polymeric matrix, graphene sheets tend to aggregate due to attractive van der Waals forces. Therefore, electrospinning requires that the graphene sheets first be dispersed in the polymer solution [93].

Composite fibers can be made mainly by two processes: mixing and spinning as shown in Fig. 7. Graphene can be incorporated during the mixing process in three different ways: solvent mixing, melt processing, and in-situ polymerization. Some researchers, after the spinning process, have tried to coat graphene onto the

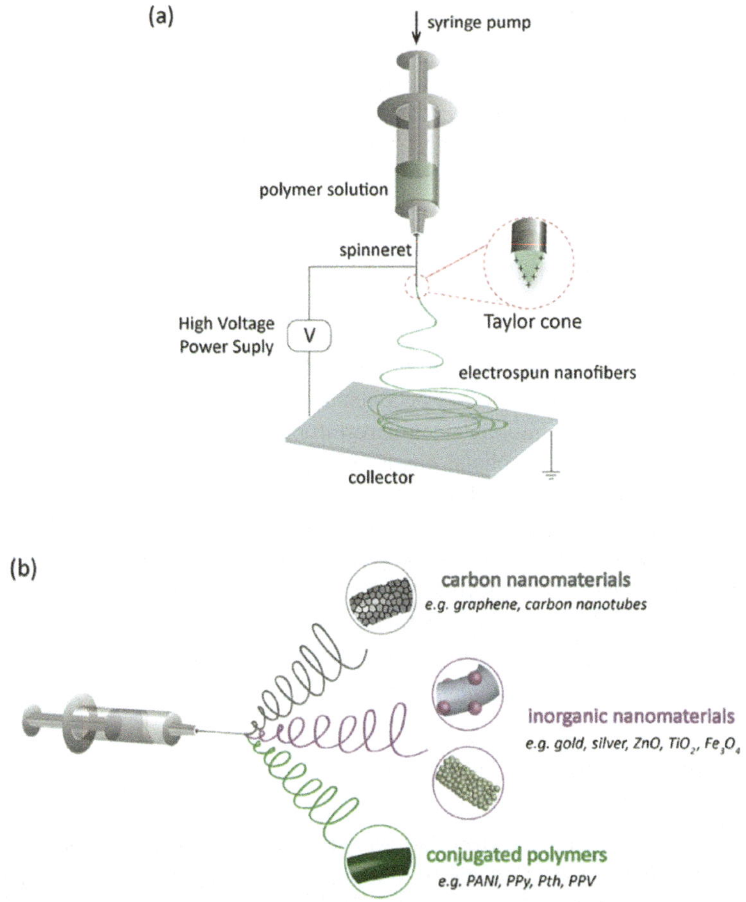

Fig. 6 Schematic diagram of **a** a typical electrospinning apparatus setup and **b** the electrospinning of multifunctional nanofibers containing different types of materials (Reproduced with permission from ref [90]. Copyright© 2017 Elsevier B.V.)

fiber surface. There are also different spinning methods to prepare graphene/polymer fibers: wet spinning, melt spinning, and electrical spinning [94].

Compared with traditional manufacturing methods (such as melt blending, solvent blending, and in-situ polymerization..., etc.), electrospinning is an ideal way to align carbon nanomaterials because of its simplicity and low cost of the processing system, the short time required to prepare continuous 1D structures and its versatility, which allows the preparation of fibers and membranes with a wide range of morphologies and materials [90–92]

Wang et al. [95] prepared the PVA/GO/TiO_2 core–shell fibers after electrospinning GO/PVA nanofibers as shown in Fig. 8. The experimental results showed that GO-doping decreased the temperature of decomposition and elongation at break

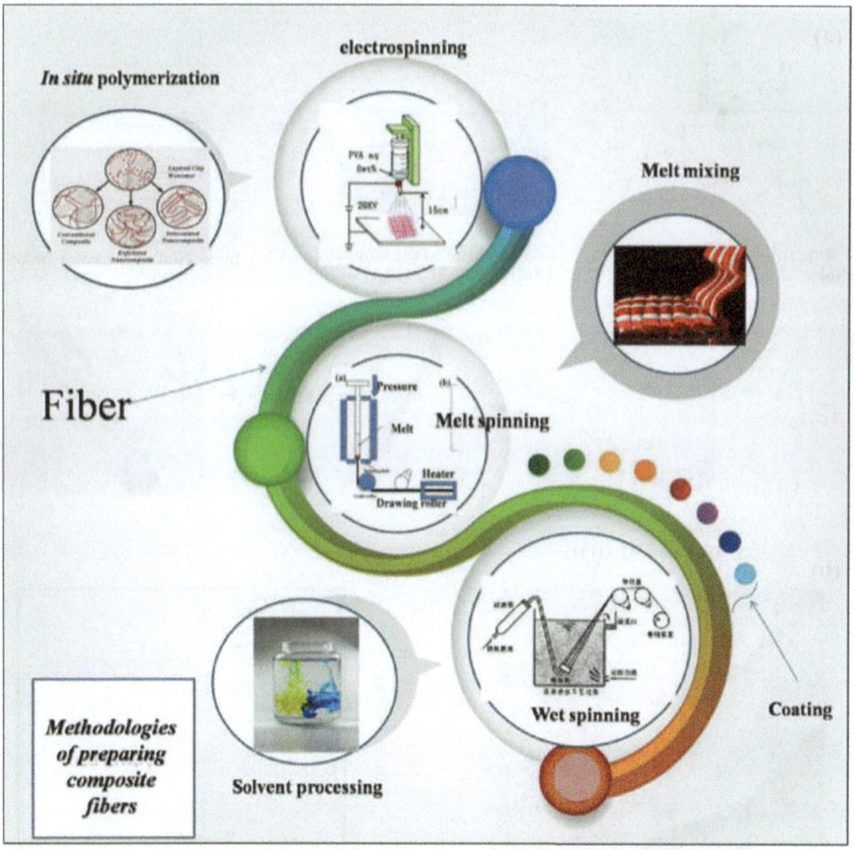

Fig. 7 Schematic illustration of the methods for the preparation of fiber composite. (Reproduced with permission from ref [94]. Copyright© 2016 Elsevier Ltd)

of PVA nanofibers, while, the degree of crystallinity and tensile strength increased Fig. 7c.The GOdoped PVA nanofibers can be used as a hard template to promote the growth of TiO_2. In addition, the PVA fibers, reinforced with both graphene and CNTs by dispersing graphene and CNTs into an aqueous PVA solution then spinning and demonstrated a strong synergistic effect by hybridization of single-walled carbon nanotubes (SWCNT) andRGO flakes [96].

Pan et al. [97] have reinforced Nylon fibers by graphene. Nylon-6 spider-wave-like nano-nets are manufactured by regulating the amount of GO in polymer solution during the electrospinning process as shown in Fig. 9a. The spider-wave-like nano-nets that comprise interlinked thick fibers (192 nm diameter) and thin (14 nm diameter) are widely distributed throughout the mat when the nylon-6 solution is reinforced with GO (Fig. 8c). The heterogeneous composite mats were composed of two nanofibers with different diameters in which pore diameter was greatly decreased by the factor of 2 compared to the pristine nylon-6 mat. The acceleration in ionization

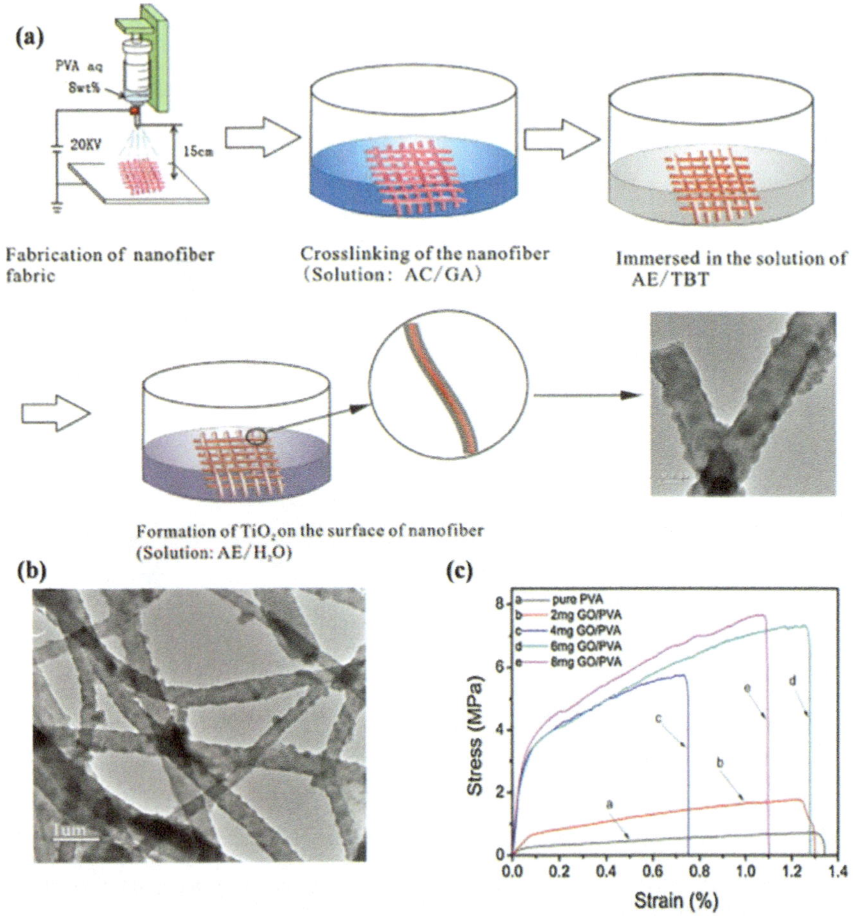

Fig. 8 The procedure for preparing of PVA/TiO_2 or PVA/GO/TiO_2 composite nanofibers through electrospinning and interface sol–gel reaction **a**. TEM images of composite nanofibers of PVA/GO/TiO_2**b**. Tensile stress–strain curves of nanofiber fabrics with various GO loading **c** .(Reproduced with permission from ref [95]. Copyright © 2014 Elsevier B.V.)

and degradation of nylon-6 (due to formic acid) solution caused by well-dispersed GO sheet as well as the formation of hydrogen bond between nylon-6 molecules and GO sheet during electrospinning are proposed as the possible mechanisms for the formation of these spider-wave-like nano-nets. Mack et al. designed a reinforced polymer composite by adding 1–4% graphite nanoplatelets into a PAN solution. Electrospinning produced composite nanofibers with average diameters of 300 nm. The composite nanofibers demonstrated a modest increase in thermal stability with increasing weight-percent graphite nanoplatelets [98].

He et al. [99] have prepared a new and special NaAlg fiber and added GO using a wet-spinning method. The NaAlg/GO fibers are nontoxic to cells which demonstrated

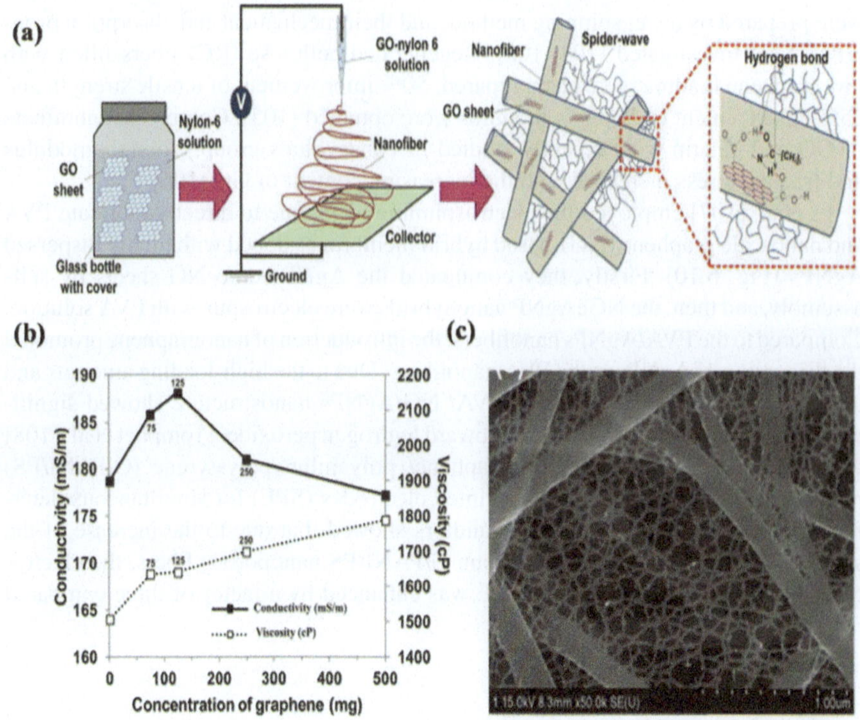

Fig. 6.9 Fabrication of spider-wave-like nano-nets and interaction of GO sheets and nylon-6 molecules by hydrogen bonding **a**. Conductivity and viscosity of GO containing nylon-6 solutions mats **b**. FE-SEM images of 125 mg GO containing nylon-6 solutions mats. (Reproduced with permission from ref [97]. Copyright © 2012 Elsevier B.V.)

the potential applications of the as-spun fibers in wound dressing materials. Li and co-workers [100], prepared a novel environmentally friendly adsorbent, calcium alginate (CA) immobilized GO fiber. Thermodynamic analysis proved that the adsorption reaction of methylene blue onto CA immobilized GO fiber was exothermic and spontaneous in nature.

Yoon et al. [101], prepared reinforced nanofibers of poly(D, L-lactic-co-glycolic acid) (PLGA) by adding 2-dimensional nanoscale fillers of GO nanosheets to PLGA nanofibers. Lamastra et al. [102], reinforced poly(e-caprolactone) (PCL) fibers with CNTs and graphene nanoplatelets (GNPs). CNTs and GNPs were accurately disentangled and dispersed in N, N-Dimethylformamide (DMF) by means of sonication, followed by electrospun with a mean diameter ranging between 1 and 2 μm. Asmatulu and co-workers [103], have fabricated PS and polyvinyl chloride (PVC) fibers incorporated TiO_2nanoparticles and graphene nanoflakes with an electrospinning technique, and then the surface morphology and super-hydrophobicity of these electrospun nanocomposite fibers were investigated. GO/chitosan composite fibers were prepared by a wet spinning method, and their mechanical and absorption properties were investigated [104, 105]. Regenerated cellulose (RC) fibers filled with low graphene loading were also prepared, 50% improvement of tensile strength and 25% enhancement of Young's modulus were obtained [105]. Composite nanofibers of GO and gelatin were recently studied in Panzavolta's group. Young's modulus and fracture stress increased with the increasing content of GO [106].

Li et al. [107] employed the electrospinning technique to directly fabricate PVA and nanoscale graphene (NG)-based hybrid membranes doped with highly dispersed AgNPs (Fig. 6.10). Firstly, they conjugated the AgNPs onto NG sheets by self-assembly, and then, the NGeAgNP nanohybrids were electrospun with PVA solution. Compared to the PVA/AgNPs nanofibers, the introduction of nanographene promoted the dispersion of AgNPs in the PVA nanofibers. Due to the high-loading amounts and uniform dispersity of AgNPs, the PVA/ NG/AgNPs nanostructure showed significantly improved catalytic properties toward hydrogen peroxide. Promphet et al. [108] used electrospinning to develop graphene/ polyaniline/polystyrene (G/PANI/PS) nanoporous fiber modified screen-printed electrodes (SPE) for simultaneous determination of Pb(II) and Cd(II). The authors showed that due to the increase of the specific surface area of the electrospun G/PANI/PS nanoporous fibers, the electrochemical sensitivity of modified SPE was enhanced by a factor of three compared

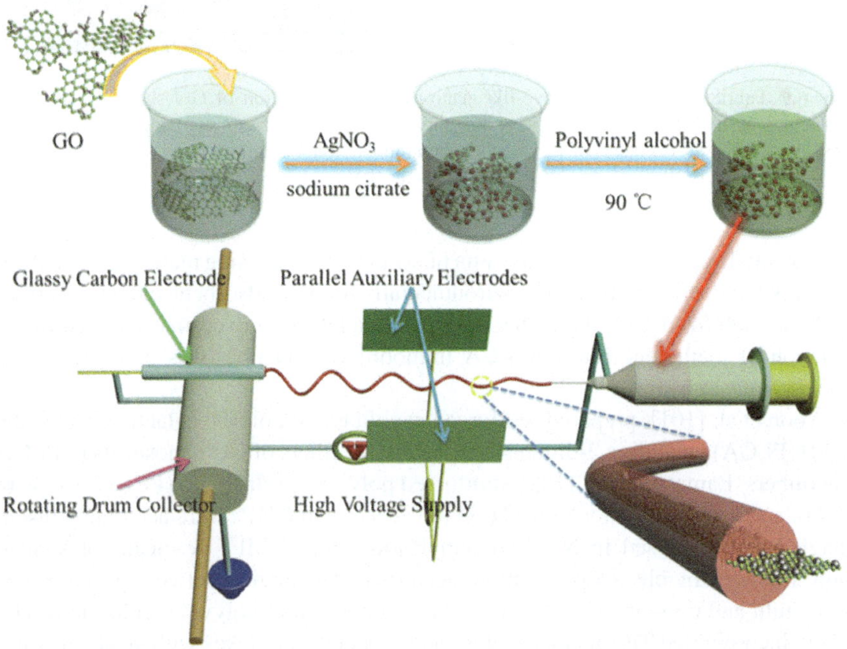

Fig. 6.10 Schematic representation for the fabrication mechanism of NG/AgNP hybrid membrane. (Reprinted with permission from ref [109]. Copyright© 2016 WILEY VCH Verlag GmbH & Co. KGaA, Weinheim)

to an unmodified SPE. Under optimal conditions, a linear relationship was found in a range of 10–500 μg L^{-1} with detection limit (S/N = 3) of 3.30 and 4.43 μg $L^{-1}$for Pb(II) and Cd(II), respectively. By simple washing step, the ESNFs-based electrode could be reused for more than ten replicates with high reproducibility for the determination of heavy metals in real river water samples.

Abideen et al. [110] developed a gas sensor platform assembling reduced graphene oxide (rGO) in ZnO ESNFs membrane (Fig. 6.10). Microstructural investigation revealed that the addition of rGO does not affect the size of the ZnO nanoparticles or nanofibers. The sensor response to 5 ppm NO_2 gas was increased by the addition of rGO to ZnO nanofibers. Overall, the rGO-loaded ZnO ESNFs showed higher sensitivity to different oxidizing and reducing gases compared to monolithic ZnO nanofibers (Fig. 6.11).

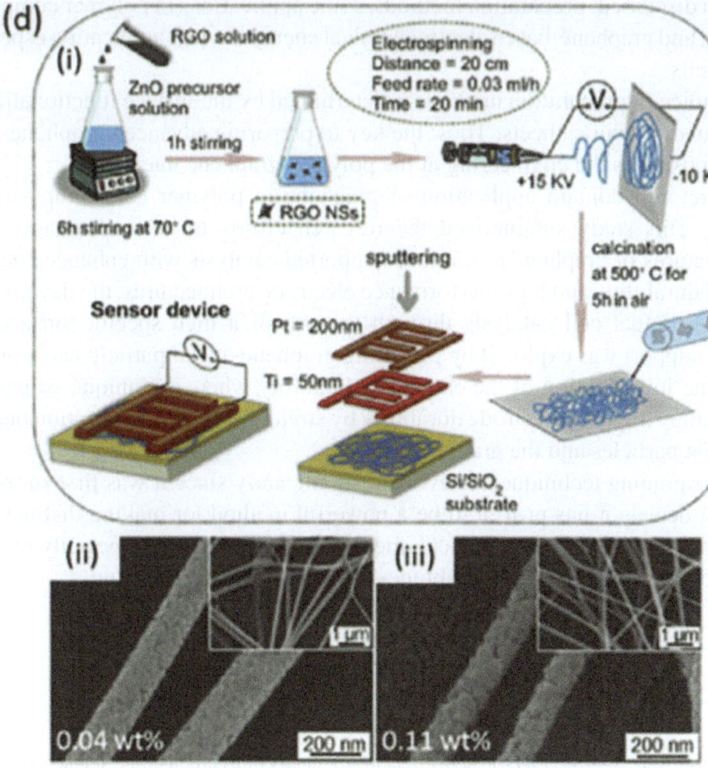

Fig. 6.11 The synthesis of rGO-loaded ZnO nanofibers via electrospinning method, (ii) and (iii) SEMimages of calcined rGO-loaded ZnO NFs. (Reprinted with permission from ref [110]. Copyright © 2015 Elsevier B.V.)

6 Conclusions

Graphene-based polymer nanocomposites have attracted a great deal of interest due to its good stability, interesting electroactivity, high capacitance, and unusual doping/de-doping chemistry, especially when combined with graphene, performance of nanocomposites is greatly improved. Therefore, graphene-based polymer nanocomposites can be widely employed in various applications.

In the field of electrochemical energy conversion and storage, graphene-based polymer nanocomposites have shown promise for applications in fuel cells, supercapacitors, and lithium-ion batteries. Due to its specific surface area and the presence of functional groups, edges, and defects, graphene-based electrodes demonstrate enhanced performance for such electrochemical energy devices.

In this book chapter, many aspects of graphene-based polymer nanocomposites have been discussed: preparation methods, some applications in polymer composites examples, and graphene-based electrochemical energy devices application especially the fuel cells.

The choice of preparation methods is determined by the surface functionalization of integrated graphitic sheets. Thus, the key to preparing advanced graphene-based nanocomposites is the engineering at the polymer-graphene interface.

The fuel cell domain application of graphene in polymer nanocomposite was reviewed. This study summarized the research efforts to take advantage of the unique features of graphene to produce supported catalysts with enhanced activity, increased durability, and high-performance electrode architectures, the development for enhanced fuel cell catalysts through the use of a high specific surface area graphene support was explored by preparing graphene–metal particle nanocomposites and the investigation of the enhanced durability where the unique structure of graphene may improve electrode durability by strengthening the interaction between the catalyst particles and the graphene support.

Electrospinning technique has evolved significantly since it was first suggested, in the last decade it has proven to be a powerful method for making distinct functional nanomaterials for non-medical, medical applications and especially in textile applications, optical devices, membranes, sensors, tissue engineering.

References

1. Novoselov, K.S., et al.: Electric field effect in atomically thin carbon films. Science **306**(5696), 666–669 (2004)
2. Geim, A.K., Novoselov, K.S.: The rise of graphene. Nat. Mater. **6**(3), 183–191 (2007)
3. Geim, A.K.: Graphene: status and prospects. Science **324**(5934), 1530–1534 (2009)
4. Stankovich, S., et al.: Graphene-based composite materials. Nature **442**(7100), 282–286 (2006)
5. Ramanathan, T., et al.: Functionalized graphene sheets for polymer nanocomposites. Nat. Nanotechnol. **3**(6), 327–331 (2008)

6. Dikin, D.A., et al.: Preparation and characterization of graphene oxide paper. Nature **448**(7152), 457–460 (2007)
7. Kim, K.S. et al.,: Large-scale pattern growt.h of graphene films for stretchable transparent electrodes. Nature **457**(7230), 706–710 (2009)
8. Reina, A., et al.: Large area, few-layer graphene films on arbitrary substrates by chemical vapor deposition. Nano Lett. **9**(1), 30–35 (2009)
9. Shi, Y.M., et al.: Work function engineering of graphene electrode via chemical doping. ACS Nano **4**(5), 2689–2694 (2010)
10. Li, X.S., et al.: Transfer of large-area graphene films for highperformance transparent conductive electrodes. Nano Lett. **9**(12), 4359–4363 (2009)
11. Bunch, J.S., et al.: Electromechanical resonators from graphene sheets. Science **315**(5811), 490–493 (2007)
12. Chen, C.Y., et al.: Performance of monolayer graphene nanomechanical resonators with electrical readout. Nat. Nanotechnol. **4**(12), 861–867 (2009)
13. Schedin, F., et al.: Detection of individual gas molecules adsorbed on graphene. Nat. Mater. **6**(9), 652–655 (2007)
14. Elias, D.C., et al.: Control of graphene's properties by reversible hydrogenation: evidence for graphane. Science **323**(5914), 610–613 (2009)
15. Shan, C.S., et al.: Direct electrochemistry of glucose oxidase and biosensing for glucose based on graphene. Anal. Chem. **81**(6), 2378–2382 (2009)
16. Balog, R., et al.: Bandgap opening in graphene induced by patterned hydrogen adsorption. Nat. Mater. **9**(4), 315–319 (2010)
17. Stoller, M.D., et al.: Graphene-based ultracapacitors. Nano Lett. **8**(10), 3498–3502 (2008)
18. Kouni, B., Belhamdi, H.: In: Surface Engineering of Graphene, pp. 231257. Springer (2019)
19. Wang, L., Lu, X., Lei, S., Song, Y.: Graphene-based polyaniline nanocomposites: preparation, properties and applications. J. Mater. Chem. A **2**, 4491 (2014)
20. Xu, Y., Wang, Y., Jiajie, L., Huang, Y., Ma, Y., Wan, X., Chen, Y.: A hybrid material of graphene and poly (3, 4-ethyldioxythiophene) with high conductivity, flexibility, and transparency. Nano Res **2**, 343–348 (2009)
21. Quan, H., Zhang, B., Zhao, Q., Yuen, R.K.K., Li, R.K.Y.: Facile preparation and thermal degradation studies of graphite nanoplatelets (GNPs) filled thermoplastic polyurethane (TPU) nanocomposites. Composites A **40**, 1506–1513 (2009)
22. Eda, G., Chhowalla, M.: Graphene-based composite thin films for electronics. Nano Lett **9**, 814–818 (2009)
23. Liang, J., Xu, Y., Huang, Y., Zhang, L., Wang, Y., Ma, Y., Li, F., Guo, T., Chen, Y.: Infrared-triggered actuators from graphene-based nanocomposites. J. Phys. Chem. **113**, 9921–9927 (2009)
24. Kim, H., Macosko, C.W.: Processing–property relationships of polycarbonate/graphene nanocomposites. Polymer **50**, 3797–3809 (2009)
25. Li, H., Pang, S., Wu, S., Feng, X., Mullen, K., Bubeck, C.: Layer-by-layer assembly and UV photoreduction of graphene-polyoxometalate composite films for electronics. J Am Chem Soc **133**, 9423–9429 (2011)
26. Cassagneau, T., Fendler, J.H.: High density rechargeable lithium-ion batteries self-assembled from graphite oxide nanoplatelets and polyelectrolytes. Adv Mater **10**, 877–881 (1998)
27. Hu, H., Wang, X., Wanga, J., Wana, L., Liu, F., Zheng, H., Chen, R., Xu, C.: Preparation and properties of graphene nanosheets-polystyrene nanocomposites via in situ emulsion polymerization. Chem Phys Lett **484**, 247–253 (2010)
28. Hu, K., Kulkarni, D.D., Choi, Vladimir V. Tsukruk.Graphene-polymer nanocomposites for structural and functional applications.Progress in Polymer Science 39 (2014) 1934–1972
29. Kalaitzidou, K., Fukushima, H., Drzal, L.T.: A new compounding method for exfoliated graphite-polypropylene nanocomposites with enhanced flexural properties and lower percolation threshold. Compos. Sci. Technol. **67**, 2045–2051 (2007)
30. Zheng, W., Lu, X., Wong, S.C.: Electrical and mechanical properties of expanded graphite-reinforced high-density polyethylene. J. Appl. Polym. Sci. **91**, 2781–2788 (2004)

31. Zhao, Y.F., Xiao, M., Wang, S.J., Ge, X.C., Meng, Y.Z.: Preparation and properties of electrically conductive PPS/expanded graphite nanocomposites. Compos. Sci. Technol. **67**, 2528–2534 (2007)
32. Du, X.S., Xiao, M., Meng, Y.Z.: Synthesis and characterization of polyaniline/graphite conducting nanocomposites. J. Polym. Sci. B Polym. Phys. **42**, 1972–1978 (2004)
33. Cho, D., Lee, S., Yang, G., Fukushima, H., Drzal, L.T.: Dynamic mechanical and thermal properties of phenylethynyl-terminated polyimide composites reinforced with expanded graphite nanoplatelets. Macromol. Mater. Eng. **290**, 179–87 (2005)
34. Kim, H., Miura, Y., Macosko, C.W.: Graphene/polyurethane nanocomposites for improved gas barrier and electrical conductivity. Chem Mater **22**, 3441–3450 (2010)
35. Steinert, B.W., Dean, D.R.: Polymer **50**(3), 898–904 (2009)
36. Bortz, D.R., Heras, E.G., Martin-Gullon, I.: Impressive fatigue life and fracture toughness improvements in graphene oxide/epoxy composites. Macromolecules **45**, 238–245 (2012)
37. Wajid, A.S., Ahmed, H.S.T., Das, S., Irin, F., Jankowski, A.F., Green, M.J.: Highperformance pristine graphene/epoxy composites with enhanced mechanical and electrical properties. Macromol. Mater. Eng. **298**, 339–347 (2012)
38. Galpaya, D., Wang, M., George, G., Motta, N., Waclawik, E., Yan, C.: Preparation of graphene oxide/epoxy nanocomposites with significantly improved mechanical properties. J. Appl. Phys. **116,** 053518 (2014)
39. Shen, X., Pei, X., Fu, S., Friedrich, K.: Significantly modified tribological performance of epoxy nanocomposites at very low graphene oxide content. Polymer **54**, 1234–1242 (2013)
40. Cooper, D.R., D'Anjou, B., Ghattamaneni, N., Harack, B., Hilke, M., Horth, A., Majlis, N., Massicotte, M., Vandsburger, L., Whiteway, E., Yu, V.: Experimental review of graphene. ISRN Condens. Mater. Phys. **2012**, 56 (2012)
41. Wei, J., Vo, T., Inam, F.: Epoxy/graphene nanocomposites—processing and properties: a review. RSC Adv. **5**, 73510–73524 (2015)
42. Sandler, J.K.W., Kirk, J.E., Kinloch, I.A., Shaffer, M.S.P., Windle, A.H.: Ultra-low electrical percolation threshold in carbon-nanotube-epoxy composites. Polymer **44**, 5893–5899 (2003)
43. Zhao, S., Chang, H., Chen, S., Cui, J., Yan, Y.: High-performance and multifunctional epoxy composites filled with epoxide-functionalized graphene. Eur. Polymer J. **84**, 300–312 (2016)
44. Atif, R., Shyha, I., Inam, F.: Mechanical, thermal, and electrical properties of graphene-epoxy nanocomposites—a review. Polymers **8**, 281 (2016)
45. Kim, C., Khan, W., Kim, D., Cho, K., Park, S.: Graphene oxide/cellulose composite using NMMO monohydrate. Carbohydr. Polym. **86**, 903–909 (2011)
46. Luong, N.D., Pahimanolis, N., Hippi, U., Korhonen, J.T., Ruokolainen, J., Johansson, L., Nam, J., Seppälä, J.: Graphene/cellulose nanocomposite paper with high electrical and mechanical performances. J. Mater. Chem. **21**, 13991–13998 (2011)
47. Weng, Z., Su, Y., Wang, D., Li, F., Du, J., Cheng, H.: Graphene-cellulose paper flexible supercapacitors. Adv. Energy Mater. **1**, 917–922 (2011)
48. Xu, M., Huang, Q., Wang, X., Sun, R.: Highly tough cellulose/graphene composite hydrogels prepared from ionic liquids. Ind. Crops Prod. **70**, 56–63 (2015)
49. Zhao, X., Zhang, Q., Chen, D., Lu, P.: Enhanced mechanical properties of graphene-based poly(vinyl alcohol) composites. Macromolecules **43**, 2357–2363 (2010)
50. Liang, J., Huang, Y., Zhang, L., Wang, Y., Ma, Y., Guo, T., Chen, Y.: Molecular-level dispersion of graphene into poly(vinyl alcohol) and effective reinforcement of their nanocomposites. Adv. Funct. Mater. **19**, 2297–2302 (2009)
51. Lee, Y.R., Raghu, A.V., Jeong, H.M., Kim, B.K.: Properties of waterborne polyurethane/functionalized graphene sheet nanocomposites prepared by an in situ method. Macromol. Chem. Phys. **210**, 1247–1254 (2009)
52. Liang, J., Xu, Y., Huang, Y., Zhang, L., Wang, Y., Ma, Y., Li, F., Guo, T., Chen, Y.: Infrared-triggered actuators from graphene-based nanocomposites. J. Phys. Chem. C. **113**, 9921–9927 (2009)
53. Zhang, H., Zheng, W., Yan, Q., Yang, Y., Wang, J., Lu, Z., Ji, G., Yu, Z.: Electrically conductive polyethylene terephthalate/graphene nanocomposites prepared by melt compounding. Polymer **51**, 1191–1196 (2010)

54. Kim, H., Macosko, C.W.: Processing-property relationships of polycarbonate/graphene composites. Polymer **50**, 3797–3809 (2009)
55. Stankovich, S., Dikin, D.A., Dommett, G.H.B., Kohlhaas, K.M., Zimney, E.J., Stach, E.A., Piner, R.D., Nguyen, S.T., Ruoff, R.S.: Graphene-based composite materials. Nature **442**, 282–286 (2006)
56. Liu, N., Luo, F., Wu, H., Liu, Y., Zhang, C., Chen, J.: One-step ionic-liquid-assisted electrochemical synthesis of ionic-liquid-functionalized graphene sheets directly from graphite. Adv. Funct. Mater. **18**, 1518–1525 (2008)
57. Wang, R., Wang, S., Chen, S., Jiang, G.: In situ polymerization approach to poly(*E*-caprolactone)-graphene oxide composites. Design. Monom. Polym. **15**(3), 303–310 (2012)
58. Chieng, B.W., Ibrahim, N.A., Wan Yunus, W.M.Z.: Optimization of tensile strength of poly(lactic acid)=graphene nanocomposites using response surface methodology. Polym.-Plast. Technol. Eng. **51**(8), 791–799 (2012)
59. Liu, Q., Zhou, X., Fan, X., Zhu, C., Yao, X., Liu, Z.: Mechanical and thermal properties of epoxy resin nanocomposites reinforced with graphene oxide. Polym.-Plast. Technol. Eng. **51**(3), 251–256 (2012)
60. Mohamadi, S., Sanjani, N.S., Mahdavi, H.: Functionalization of graphene sheets via chemically grafting of PMMA chains through in situ polymerization. J. Macromol. Sci. Pt. A **48**(8), 577–582 (2011)
61. Liang, J., Huang, L., Li, N., Huang, Y., Wu, Y., Fang, S., Oh, J., Kozlov, M., Ma, Y., Li, F., Baughman, R., Chen, Y.: Electromechanical actuator with controllable motion, fast response rate, and highfrequency resonance based on graphene and polydiacetylene. ACS Nano **6**(5), 4508–4519 (2012)
62. Pan, B., Zhao, J., Zhang, Y., Zhang, Y.: Wear performance and mechanisms of polyphenylene sulfide=polytetrafluoroethylene wax composite coatings reinforced by graphene. J. Macromol. Sci. Pt. B Phys. **51**(6), 1218–1227 (2012)
63. Pang, H., Bao, Y., Lei, J., Tang, J.H., Ji, X., Zhang, W.Q., Chen, C.: Segregated conductive ultrahigh-molecular-weight polyethylene composites containing high-density polyethylene as carrier polymer of graphene nanosheets. Polym.-Plast. Technol. Eng. **51**(14), 1483–1486 (2012)
64. Hou, J., Shao, Y., Ellis, M.W., Moore, R.B., Yi, B.: Graphene-based electrochemical energy conversion and storage: fuel cells, supercapacitors and lithium ion batteries. Phys. Chem. Chem. Phys. **13**, 15384–15402 (2011)
65. Tang, L.H., et al.: Preparation, structure, and electrochemical properties of reduced graphene sheet films. Adv. Funct. Mater. **19**(17), 2782–2789 (2009)
66. Zhou, M., Zhai, Y.M., Dong, S.J.: Electrochemical sensing and biosensing platform based on chemically reduced graphene oxide. Anal. Chem. **81**(14), 5603–5613 (2009)
67. McCreery, R.L.: Advanced carbon electrode materials for molecular electrochemistry. Chem. Rev. **108**(7), 2646–2687 (2008)
68. Pumera, M.: Electrochemistry of graphene: new horizons for sensing and energy storage. Chem. Rec. **9**(4), 211–223 (2009)
69. Steele, B.C.H., Heinzel, A.: Materials for fuel-cell technologies. Nature **414**(6861), 345–352 (2001)
70. Borup, R., et al.: Scientific aspects of polymer electrolyte fuel cell durability and degradation. Chem. Rev. **107**(10), 3904–3951 (2007)
71. Shao, Y.Y., Yin, G.P., Gao, Y.Z.: Understanding and approaches for the durability issues of Pt-based catalysts for PEM fuel cell. J. Power Sources **171**(2), 558–566 (2007)
72. Xu, C., Wang, X., Zhu, J.W.: Graphene-metal particle nanocomposites. J. Phys. Chem. C **112**(50), 19841–19845 (2008)
73. Li, Y.M., Tang, L.H., Li, J.H.: Preparation and electrochemical performance for methanol oxidation of Pt/graphene nanocomposites. Electrochem. Commun. **11**(4), 846–849 (2009)
74. Sharma, S., et al.: Rapid microwave synthesis of CO tolerant reduced graphene oxide-supported platinum electrocatalysts for oxidation of methanol. J. Phys. Chem. C **114**(45), 19459–19466 (2010)

75. Hou, J.B., et al.: Electrochemical impedance investigation of proton exchange membrane fuel cells experienced subzero temperature. J. Power Sources **171**(2), 610–616 (2007)
76. Seger, B., Kamat, P.V.: Electrocatalytically active graphene platinum nanocomposites. role of 2-D carbon support in PEM fuel cells. J. Phys. Chem. C. **113**(19), 7990–7995 (2009)
77. Dong, L.F., et al.: Graphene-supported platinum and platinum ruthenium nanoparticles with high electrocatalytic activity for methanol and ethanol oxidation. Carbon **48**(3), 781–787 (2010)
78. Bong, S., et al.: Graphene supported electrocatalysts for methanol oxidation. Electrochem. Commun. **12**(1), 129–131 (2010)
79. Kou, R., et al.: Enhanced activity and stability of Pt catalysts on functionalized graphene sheets for electrocatalytic oxygen reduction. Electrochem. Commun. **11**(5), 954–957 (2009)
80. Yoo, E., et al.: Enhanced electrocatalytic activity of Pt subnanoclusters on graphene nanosheet surface. Nano Lett. **9**(6), 2255–2259 (2009)
81. Yumura, T., et al.: The use of nanometer-sized hydrographene species for support material for fuel cell electrode catalysts: a theoretical proposal. Phys. Chem. Chem. Phys. **11**(37), 8275–8284 (2009)
82. Shao, Y.Y., et al.: Highly durable graphene nanoplatelets supported Pt nanocatalysts for oxygen reduction. J. Power Sources **195**(15), 4600–4605 (2010)
83. Groves, M.N., et al.: Improving platinum catalyst binding energy to graphene through nitrogen doping. Chem. Phys. Lett. **481**(4–6), 214–219 (2009)
84. Jafri, R.I., et al.: Nanostructured Pt dispersed on graphenemultiwalled carbon nanotube hybrid nanomaterials as electrocatalyst for PEMFC. J. Electrochem. Soc. **157**(6), B874–B879 (2010)
85. Guo, S.J., Dong, S.J., Wang, E.W.: Three-dimensional Pt-on-Pd bimetallic nanodendrites supported on graphene nanosheet: Facile synthesis and used as an advanced nano electron catalyst for methanol oxidation. ACS Nano **4**(1), 547–555 (2010)
86. Hou, J. et al.: Graphene-based electrochemical energy conversion and storage: fuel cells, supercapacitors and lithium ion batteries. Phys. Chem. Chem. Phys. 15384 (2011)
87. Qu, L.T., et al.: Nitrogen-doped graphene as efficient metal-free electrocatalyst for oxygen reduction in fuel cells. ACS Nano **4**(3), 1321–1326 (2010)
88. Gong, K.P., et al.: Nitrogen-doped carbon nanotube arrays with high electrocatalytic activity for oxygen reduction. Science **323**(5915), 760–764 (2009)
89. Shao, Y.Y., et al.: Nitrogen-doped graphene and its electrochemical applications. J. Mater. Chem. **20**(35), 7491–7496 (2010)
90. Mercante, L.A., et al.: Electrospinning-based (bio)sensors for food and agricultural applications: a review. TrAC Trends Anal. Chem. **1**(91), 91–103 (2017)
91. Navarro-Pardo, F. et al.: Carbon nanotube and graphene-based polyamide electrospun nanocomposites: a review. J. Nanomater. (2016)
92. Guo, Y., et al.: Significantly enhanced and precisely modeled thermal conductivity in polyimide nanocomposites with chemically modified graphene via in situ polymerization and electrospinning-hot press technology. J. Mater. Chem. C. **6**(12), 3004–3015 (2018)
93. Das, S., et al.: Electrospinning of polymer nanofibers loaded with noncovalently functionalized graphene. J. Appl. Polym. Sci. **128**(6), 4040–4046 (2013)
94. Ji, X., et al.: Review of functionalization, structure and properties of graphene/polymer composite fibers. Compos. A Appl. Sci. Manuf. **1**(87), 29–45 (2016)
95. Wang, B., et al.: Fabrication of PVA/graphene oxide/TiO2 composite nanofibers through electrospinning and interface sol–gel reaction: Effect of graphene oxide on PVA nanofibers and growth of TiO_2. Colloids Surf. A **5**(457), 318–325 (2014)
96. Shin, M.K., et al.: Synergistic toughening of composite fibres by self-alignment of reduced graphene oxide and carbon nanotubes. Nat. Commun. **3**(1), 1–8 (2012)
97. Pant, H.R., et al.: Bimodal fiber diameter distributed graphene oxide/nylon-6 composite nanofibrous mats via electrospinning. Colloids Surf., A **5**(407), 121–125 (2012)
98. Mack, J.J., et al.: Graphite nanoplatelet reinforcement of electrospun polyacrylonitrile nanofibers. Adv. Mater. **17**(1), 77–80 (2005)

99. He, Y., et al.: Alginate/graphene oxide fibers with enhanced mechanical strength prepared by wet spinning. Carbohyd. Polym. **88**(3), 1100–1108 (2012)
100. Li, Y., et al.: Methylene blue adsorption on graphene oxide/calcium alginate composites. Carbohyd. Polym. **95**(1), 501–507 (2013)
101. Yoon, O.J., et al.: Nanocomposite nanofibers of poly (D, L-lactic-co-glycolic acid) and graphene oxide nanosheets. Compos. A Appl. Sci. Manuf. **42**(12), 1978–1984 (2011)
102. Lamastra, F.R., et al.: Poly (ε-caprolactone) reinforced with fibres of poly (methyl methacrylate) loaded with multiwall carbon nanotubes or graphene nanoplatelets. Chem. Eng. J. **1**(195), 140–148 (2012)
103. Asmatulu, R., Ceylan, M., Nuraje, N.: Study of superhydrophobic electrospun nanocomposite fibers for energy systems. Langmuir **27**(2), 504–507 (2011)
104. Li, Y., et al.: Mechanical and dye adsorption properties of graphene oxide/chitosan composite fibers prepared by wet spinning. Carbohyd. Polym. **15**(102), 755–761 (2014)
105. Tian, M., et al.: Enhanced mechanical and thermal properties of regenerated cellulose/graphene composite fibers. Carbohyd. Polym. **13**(111), 456–462 (2014)
106. Panzavolta, S., et al.: Structural reinforcement and failure analysis in composite nanofibers of graphene oxide and gelatin. Carbon **1**(78), 566–577 (2014)
107. Li, D., et al.: Nanofibers of conjugated polymers prepared by electrospinning with a two-capillary spinneret. Adv. Mater. **16**(22), 2062–2066 (2004)
108. Promphet, N., et al.: An electrochemical sensor based on graphene/polyaniline/ polystyrene nanoporous fibers modified electrode for simultaneous determination of lead and cadmium. Sens. Actuators B Chem. **1**(207), 526–534 (2015)
109. Li, Y., et al.: Nanoscale graphene doped with highly dispersed silver nanoparticles: quick synthesis, facile fabrication of 3D membrane-modified electrode, and super performance for electrochemical sensing. Adv. Func. Mater. **26**(13), 2122–2134 (2016)
110. Abideen, Z.U., et al.: Excellent gas detection of ZnO nanofibers by loading with reduced graphene oxide nanosheets. Sens. Actuators B Chem. **31**(221), 1499–1507 (2015)

The manufacturer's authorised representative in the EU is Springer Nature Customer Service Centre GmbH, Europaplatz 3, 69115 Heidelberg, Germany. If you have any concerns regarding our products, please contact ProductSafety@springernature.com

Printed and bound by CPI Group (UK) Ltd, Croydon, CR0 4YY
30/09/2026
02225665-0016